Modern Data Analysis

Academic Press Rapid Manuscript Reproduction

Proceedings of a Workshop on Modern Data Analysis
sponsored by the United States Army Research Office
held in Raleigh, North Carolina, June 2–4, 1980.

MODERN DATA ANALYSIS

Edited by

ROBERT L. LAUNER

Mathematics Division
U.S. Army Research Office
Research Triangle Park, North Carolina

ANDREW F. SIEGEL

Department of Statistics
Princeton University
Princeton, New Jersey

1982

ACADEMIC PRESS
A Subsidiary of Harcourt Brace Jovanovich, Publishers
New York London
Paris San Diego San Francisco São Paulo Sydney Tokyo Toronto

ACADEMIC PRESS, INC.
111 Fifth Avenue, New York, New York 10003

United Kingdom Edition published by
ACADEMIC PRESS, INC. (LONDON) LTD.
24/28 Oval Road, London NW1 7DX

Library of Congress Cataloging in Publication Data
Main entry under title:

Modern data analysis.

Papers presented at a Workshop on Modern Data Analysis, organized by the Mathematics Division of the U.S. Army Research Office, and held in Raleigh, N.C., June 2-4, 1980.
Includes indexes.
Contents: Introduction to styles of data analysis techniques / John W. Tukey -- Some multiple Q-Q plotting procedures / Nicholas P. Jewell -- A reader's guide to smoothing scatterplots and graphical methods for regression / William S. Cleveland -- [etc.]
1. Mathematical statistics--Congresses. I. Launer, Robert L. II. Siegel, Andrew F. III. United States. Army Research Office. Mathematics Division. IV. Workshop on Modern Data Analysis (1980: Raleigh, N.C.)
QA276.A1M63 001.64 82-4037
ISBN 0-12-438180-4 AACR2

PRINTED IN THE UNITED STATES OF AMERICA

82 83 84 85 9 8 7 6 5 4 3 2 1

CONTENTS

CONTRIBUTORS

Numbers in parentheses indicate the pages on which the authors' contributions begin.

William S. Cleveland (37), *Bell Laboratories, Murray Hill, New Jersey*

Christopher Cox (45), *Department of Statistics, Division of Biostatistics, University of Rochester, Rochester, New York*

Jerome H. Friedman (123), *Stanford Linear Accelerator Center, Stanford, California*

K. Ruben Gabriel (45), *Department of Statistics and Division of Biostatistics, University of Rochester, Rochester, New York*

Nicholas P. Jewell (13), *Department of Statistics, Princeton University, Princeton, New Jersey*

Joseph W. McKean (171), *Department of Mathematics, Western Michigan University, Kalamazoo, Michigan*

Ronald M. Schrader (171), *Department of Mathematics and Statistics, The University of New Mexico, Albuquerque, New Mexico*

Andrew F. Siegel (103), *Department of Statistics, Princeton University, Princeton, New Jersey*

Werner Stuetzle (123), *Department of Statistics, Stanford University; Stanford Linear Accelerator Center, Stanford, California*

John W. Tukey (1, 83), *Research-Communications Principles Division, Bell Laboratories, Murray Hill, New Jersey; Department of Statistics, Princeton University, Princeton, New Jersey*

Roy E. Welsch (149), *Sloan School of Management, Massachusetts Institute of Technology, Cambridge, Massachusetts*

PREFACE

A variety of techniques which trace their origins back to many different disciplines are loosely grouped together under the heading "data analysis." What they share is the ability to work well in the task of finding useful structure in complicated collections of recorded information. They are designed to help researchers separate important features from randomness and to draw attention to specific aspects of potential interest that might otherwise be lost in a morass of supportive detail. Although some of these methods were originally created in order to solve specific problems, techniques of very general applicability have resulted. It is certainly to the benefit of scientists and statisticians to be aware of and to share theories and methods of data analysis.

Despite its usefulness, the field of data analysis is only now being accorded the acceptance and recognition it deserves as a serious branch of the science of statistics. This may be due to the necessarily fragmented history of a subject whose original contributions came in part from scientists in scattered fields. Another possible explanation for the only recent emergence of this field might be that some techniques are informal and, although they work well in practice, their precise mathematical properties are not yet known. Thus two important aspects in the study of data analysis are the creation of new methods and the derivation of the properties of existing methods.

The recent increase in awareness and acceptance of the field of data analysis is due in a large part to the efforts of John W. Tukey. Professor Tukey has practiced data analysis for many years and is responsible for the creation of a variety of new and useful methods. His 1977 book "Exploratory Data Analysis" represents a large collection of both the philosophy and the methods of analyzing data, and could be viewed as formally marking the beginning of the movement. His 1977 book with F. Mosteller, "Data Analysis and Regression," extends many of these ideas and provides a link with the more confirmatory side of data analysis. In addition to these books, Professor Tukey has expended a good deal of energy in helping the field by lecturing at scientific meetings, teaching short courses, and directing research in related areas.

How do the methods of data analysis differ from the more classical statistical techniques? A very significant difference is that traditional methods often require that a specific and often restrictive set of assumptions hold. Should the assumptions fail, the conclusions are not guaranteed to be valid, and serious errors can result without warning. In contrast, exploratory and graphic data analytic methods are designed to help the researcher detect many different types of structures. Thus many good data analytic methods are robust and still work well under a variety of underlying models, especially in the presence of outliers and errors in the data. But perhaps the largest difference between classical statistics and modern data analysis is in the philosophy behind the methods. Many classical methods are designed with one model and one question in mind, and the methods are optimized accordingly. Good data analytic methods are designed with the unexpected in mind, so that potentially crucial facets of the data will not be overlooked.

The flow of material in this volume roughly proceeds from general and exploratory to specific and confirmatory. We begin with an introduction to the styles of data analysis, leading into several papers featuring graphical methods. These are followed by a series of contributions relating to the recognition of mathematical form and physical structure. The final papers are closely concerned with the development of formal theories with application to robustness in regression and the linear model.

These papers were presented at the workshop on Modern Data Analysis in June 1980 in Raleigh, North Carolina, organized by the Mathematics Division of the U.S. Army Research Office.

ABSTRACTS

INTRODUCTION TO STYLES OF DATA ANALYSIS TECHNIQUES

John W. Tukey

We are not used to thinking about data analysis techniques in terms of style. We are not familiar with a good supply of names or acronyms for either the broad purposes of the techniques or the other important "coordinates" in whose terms such techniques can be usefully described. As a result both writer and reader have an unusually difficult task. The development sequence begins with a sketch of three pairs of coordinates. The first pair, "stochastic background" and "stringency," seem to deserve treatment together, many instances falling under one of eight rubrics. Another pair "character" and "flexibility" also go together with six combinations worth emphasis. The combination (interactive) of these two pairs of coordinates is then described. Next we notice that "data structure" is wisely interpreted as covering more than the externals of the data, going on to a brief historical setting for modern robust/resistant techniques.

SOME MULTIPLE Q-Q PLOTTING PROCEDURES

Nicholas P. Jewell

This paper is concerned with some extensions to the idea of a quantile-quantile (Q-Q) plot that is commonly used by statisticians for a variety of purposes. Both single and multiple Q-Q plots are considered. Particular attention is paid to problems involving extreme-value data and to the study of the behavior of sample averages.

A READER'S GUIDE TO SMOOTHING SCATTERPLOTS AND GRAPHICAL METHODS FOR REGRESSION

William S. Cleveland

Comments about smoothing scatterplots and graphical methods for regression are made and pointers to literature relevant to these comments are given.

SOME COMPARISONS OF BIPLOT DISPLAY AND PENCIL-AND-PAPER EXPLORATORY DATA ANALYSIS METHODS

Christopher Cox and K. Ruben Gabriel

This paper uses a number of data sets from Tukey's "Exploratory Data Analysis" to compare their inspection and analysis by biplot display with the exploratory data analysis given by Tukey. The use of biplots for display of two and three way tables is described and the methods of diagnosing models are explained. The illustrations suggest that biplot diagnoses usually result in similar models (additive, multiplicative, degree-of-freedom-for-non-additivity, etc.) to those brought out by pencil-and-paper exploratory data analysis techniques—but biplot diagnostics seem much faster and more immediate.

THE USE OF SMELTING IN GUIDING RE-EXPRESSION

John W. Tukey

Most frameworks, whether or not statistical models, used in data analysis involve some type of functional behavior. Thus it is important to have effective techniques of asking the data what sort of functional behavior we should have in our framework for handling a specific body (or kind) of data. Some sort of smoothing process is essential here; one that is robust/resistant and provides specially smooth input to a recognizer of functional form.

Smoothing is usually thought of as value change, but it can also be done by eliminating less typical points and keeping more typical ones. *Smelting* is a specific class of techniques for smoothing by excision, in which the qualitative nature of the series is used to tell us which (x, y) pairs to keep and which to set aside.

Combined with a good selection of diagnostic plots, smelting offers the best route we have today toward functional form recognition. Done reasonably, we can go far toward the use of functional forms invertible in closed form, avoiding, for example, dangerous polynomials.

GEOMETRIC DATA ANALYSIS: AN INTERACTIVE GRAPHICS PROGRAM FOR SHAPE COMPARISON

Andrew F. Siegel

Two shapes, each consisting of n homologous points, can be rotated, scaled, and translated to obtain a close fit to each other by several methods. An interactive graphical computer program is presented here that implements two methods: least squares and repeated medians, a robust method. Examples are given and the use of the system is discussed.

PROJECTION PURSUIT METHODS FOR DATA ANALYSIS

Jerome H. Friedman and Werner Stuetzle

Projection pursuit methods iteratively construct a model for structure in multivariate data, based on suitably chosen lower dimensional projections. At each step of the iteration, the model is updated to agree with the data in the corresponding projection. Projections can be chosen either by numerical optimization (automatic projection pursuit) or interactively by a user at a computer graphics terminal (manual projection pursuit). The projection pursuit paradigm has been applied to clustering, regression, classification, and density estimation.

INFLUENCE FUNCTIONS AND REGRESSION DIAGNOSTICS

Roy E. Welsch

Influential-data diagnostics are becoming an accepted part of data analysis. In this paper we show how these diagnostic techniques are connected with the ideas of qualitative robustness described by Hampel and the concept of bounded-influence regression as developed by Krasker and Welsch. Asymptotic influence functions are discussed, and the identification of influential subsets of data points is considered.

THE USE AND INTERPRETATION OF ROBUST ANALYSIS OF VARIANCE

Joseph W. McKean and Ronald M. Schrader

Robust analysis of variance procedures are discussed for the general non-full-rank linear model. These procedures are described in terms of their mathematical structure, demonstrating that the analysis has the same uses and interpretation as classical analysis of variance. This structure also leads to efficient computational algorithms. Necessary standardizing constants for the test statistics are motivated by consideration of likelihood ratio tests. An example of an experimental design illustrates the similarity between the robust and classical analyses, emphasizing the advantages of the robust method. Some Monte Carlo results attest to the validity of the robust methods for the example.

INTRODUCTION TO STYLES OF DATA ANALYSIS TECHNIQUES

John W. Tukey

Research-Communications Principles Division
Bell Laboratories
Murray Hill, New Jersey
and
Department of Statistics
Princeton University*
Princeton, New Jersey

We are not used to thinking about data analysis techniques in terms of style. We are not familiar with a good supply of names or acronyms for either the broad purposes of the techniques or the other important "coordinates" in whose terms such techniques can be usefully described. As a result both writer and reader have an unusually difficult task.

One way to, I hope, ease that task is to offer readers two different ways to read what follows: *either* Sections I to VI followed by Sections VII and VIII, thus developing concepts before their exemplification, *or* first Sections VII and VIII, followed by Sections I to VI, thus illustrating the concepts before defining them.

The development sequence begins [Section I] with a sketch of three pairs of coordinates. The first pair, "stochastic background" and "stringency", seem to deserve treatment together [Section II], many instances falling under one of 8 rubrics. Another pair "character" and "flexibility" also go together [Section III] with 6 combinations worth emphasis. Section IV then describes the combination (interactive) of these two pairs of coordinates. Next we notice [Section V] that "data structure" is wisely interpreted as covering more than the externals of the data, going on [Section VI] to a brief historical setting for modern robust/resistant techniques.

The illustration sequence begins [Section VII] with a brief account of the more important classes of data-handling components -- ADEs, OCONs, DDAPs, OUTs, CDAPs and SDAPs. Section VIII discusses how more or less familiar techniques, when used to analyze 10 observations on each of 7 quantities, fit into the classification set up in Section VI and, in part, into the coordinates described earlier.

*Prepared in part in connection with research at Princeton University sponsored by the Army Research Office (Durham).

ISBN 0-12-438180-4

I. COORDINATES FOR USES OF DATs

If we are to think about uses of data analysis techniques (DATs), we need to have several kinds of coordinates in mind. These are conveniently grouped as follows:

styles {
- *stochastic background*
- *indication, conclusion, etc.*

data structure {
- *formal arrangement*
- *type of phenomena*

specifiers {
- *output wanted*
- *algorithms used*

Any or all may be important. Of the six, the 3rd, 5th and 6th are, relatively at least, well understood. What we might need to discuss then, will be the 1st, 2nd, and, in less detail, the 4th.

Some readers may want to begin with Section VII, reading to the end before returning to this point. Others will prefer to read the sections in order of their numbers.

II. STOCHASTIC BACKGROUND AND STRINGENCY COMBINED: THE FIRST STYLE COORDINATE

It would have been possible for techniques to be common with any combination of stochastic background and stringency.

Here "stringency" is a deliberately vague term (cp. Mosteller and Tukey 1977, pp. 17ff) covering "efficiency", "power", "minimum variance" and the like. In our current world, however, only a few combinations are at all common, namely the eight in exhibit 1.

Two comments need to be made about this coordinate:

- where a DAT (data analysis technique) belongs may depend on the circumstances where it was invented -- and it may depend on how it is thought about. Thus moment-matching requires little if any formal background for its invention (and development), but *is* sometimes of high narrow stringency against an overutopian background.
- one reason for the absence of "nonparametric, high" is that we have not found any good way to seek out such a behavior.

exhibit 1

The 8 common combinations of stochastic background and stringency

Stochastic background	Narrow stringency*	Broad stringency*	Examples** (say for $n \geq 8$)
No formal	(inapplicable)	(inapplicable)	midhinge
Overutopian	Not much	Dubious	s^2 and mean of random subsample of 3
Overutopian	High or nearly so	Dubious	mean and s^2
"Nonparametric"	Unknown	Unknown	median and sign test for a random subsample
"Nonparametric"	Some	Dubious	midhinge and hingespread
"Nonparametric"	Some	Some	—
Robust/resistant	Medium	Medium	midmean and midspread
Robust/resistant	High	High	biweight

*"Stringency" here means "degree of success in wringing out all the information that is there". It is narrow if assessed for a narrowly specified situation, as for instance, for samples from a Gaussian distribution. It is broad if assessed against each and any of a broad set of situations.

**Most unfamiliar terms are defined in either Tukey 1977, Mosteller and Tukey 1977, or both.

Illustrations

A few brief illustrations may help us. Let us consider:

- Using Student's t is a matter of *critical* data analysis. This is usually thought of as deriving from a tight Gaussian specification via sufficient statistics (which would make its stochastic background "overutopian", its narrow stringency "high", and its broad stringency "dubious"), but, since the paper of Pitman (1937), using Student's t can also be thought of -- in the two-sample case, at least -- as almost nonparametric in stochastic background but still, of course, with broad stringency, dubious.
- Looking hard at a sample median (unaccompanied); this has to be a matter of *exploratory* data analysis, probably with no formal stochastic background for small batches, however, medians also come from a robust/resistant stochastic background; the broad stringency would vary with batch size (#=3 or 4, evermore relatively high stringency; #=5 or 6, today relatively high stringency. $n = 7$ up, moderate stringency.)
- Looking hard at a modern (say 6-biweight) estimate of center unaccompanied by assessment of width; this still has also to be a matter of *exploratory* data analysis, but the stochastic background is almost surely robust/resistant, and the stringency is high (at least for $n \geq 7$ or 8).

III. CHARACTER AND FLEXIBILITY COMBINED: THE SECOND STYLE COORDINATE

Here we find character and flexibility even more closely related to one another (than stochastic background and stringency were). The six alternatives of exhibit 2 cover all that is today common.

exhibit 2

6 common combinations of character and flexibility

Tag	Character	Flexibility
RDDA	rigidly descriptive (with no exploration)	negligible
EDA	truly exploratory	large
OCDA→E	"Overlapping-critical" used for exploratory purposes	yes
OCDA→C	"Overlapping-critical" used for confirmatory purposes	yes
SCDA	{"simple confirmatory" / "separate critical"}	not here
CCDA	"careful confirmatory"	eliminated

By rigidly descriptive data analysis (with no exploration) -- RDDA -- we understand a wholly descriptive analysis -- which had been prescribed, in complete detail, before the data had been seen. (We can be surest of this last, of course, if the prescription is completed before the data is obtained.) Such an analysis offers a sample mean and a sample variance as estimators of a population mean and population variance, but could (in a particular instance) eschew standard error (estimated standard deviation) for the sample mean (to say nothing of related tests of significance or confidence intervals).

Such a rigid technique has only one virtue -- beyond whatever general usefulness a good description might provide (or the actual description does provide) -- knowledge of the specific idiosyncrasies of the data cannot have influenced the choice of a (prescribed) analysis. This is only very occasionally useful.

Exploratory data analysis, EDA, calls for a relatively free hand in exploring the data, together with dual obligations:

- to look for all plausible alternatives or oddities -- and a few implausible ones, (graphic techniques can be most helpful here) and
- to remove each appearance that seems large enough to be meaningful -- ordinarily by some form of fitting, adjustment, or standardization (for the last, see e.g. Mosteller and Tukey 1977, Chapter 11 for the most common forms of standardization) -- so that what remains, the residuals, can be examined for further appearances.

OCDA -- an overlapping of exploration and critical assessment of the *same* data -- is what we often do -- and often must do, if we are to meet our responsibilities -- first explore and then use critical/confirmatory techniques in ways suggested by, or chosen with the aid of, the results of our exploration. Our critical/confirmatory techniques -- whether stating a bare standard error, doing a test of significance, making one or more tests of directionality, or making a confidence statement (simple or compound) -- will not have quite the same statistical properties when their choice is guided by exploration -- is part of an OCDA, rather than having this choice prescribed in some data-unaffected way (when our critical/confirmatory technique would be part of an SCDA). OCDAs lose a real sort of definiteness, since we cannot be as sure of our technique's statistical properties, but the gains from having explored, and used the results of exploration as guidance, outweigh such losses more often than not.

SCDA, -- either "simple **c**onfirmatory" or "separately **c**ritical" **d**ata **a**nalysis -- is any form of critical/confirmatory data analysis whose choice was uninfluenced by the data to which it is being applied. (We expect and urge, the use of explorations of *similar* data as part of the advance choice of prescription.) We notice that such critical/confirmatory techniques may be part of a larger exploration (OCDA→E), but are more likely to be used for purely confirmatory purposes.

Finally, we need to plan, in the most crucial situations, such as weather modification trials and clinical trials of drugs or curative procedures, for not only advance prescription of the analysis, but also the greatest care in understanding and monitoring measurement techniques of all kinds, etc. When we have done all that insight and expertise can suggest, we will have really done careful confirmatory data analysis (CCDA).

(Foı some further discussion of these concepts, see Mallows and Tukey 1981).

IV. HOW THE STYLE COORDINATES COMBINE

We now ask how *background and stringency* combines with *character and flexibility*. Roughly, I believe, as in exhibit 3.

exhibit 3

Viable, avoidable and inconceivable combinations of character and flexibility with stochastic background and (broad) stringency

		None	Overutopian		Non-parametric		Rob/res	
			Not Much	High	Unknown	Some	Median	High
r	RDDA	→→	→→	→	→	vv	vv	vvv
fff	EDA	→	→→	→	→	vv	vv	vvv
ff	OCDA→E	→	→→	→	→	v	vv	vvv
ff	OCDA→C	x	x	→→	→→→	v	v	vvv
r	SCDA	x	x	→→	→→→	v	→	vvv
rr	CCDA	x	x	→→→	x	→	→	vvv

where the coding is:

r, rr,...	=	for increasing rigidity,
f, ff,...	=	increasing flexibility,
and		
x	=	no, never
→	=	prefer to move to right, if possible
→→	=	strongly prefer to move to right, if possible
→→→	=	must move to right, if possible
v	=	might do, but...
vv	=	probably okay
vvv	=	quite okay

(Rows may not be choosers!)

V. A FEW WORDS ABOUT THE DATA STRUCTURE COORDINATES

"Data structure" usually refers, in past accounts, to the formal structure of the data, as for instance:

- a batch (perhaps "a sample")
- a two-way table
- a Greco-Latin square
- a time series
- 16 synchronized time series

Just describing this coordinate is not enough, however. "A time series" may be "like" microseisms or ocean-wave heights. If it is, it probably deserves analysis "on the frequency side". Not only will we use the FFT (the Fast Fourier Transform), but we will make our interpretations mainly on the frequency side.

Another time series may be "like" distant earthquakes, tsunamis, or geophysical prospectors' seismograms. If it is, it probably deserves analysis "on the time side". We are likely to use the FFT an even number of times, and make our interpretations mainly on the "time" (perhaps "space") side.

Still another time series may be "like" conventional economic series. If so, it may help to divide it into a sum of 3 parts — conventionally called "trend-cycle", "seasonal" and "irregular". (We might use "X11" or, better, "SABL" for this purpose -- see Cleveland et al 1979 and their references -- perhaps with Dagum trimmings ("X11-?" or better).

Yet another time series might be "like" the enrollment in first Yale College and then Yale University over the last two centuries (more or less). If so, we probably don't yet know how to analyze it, although new suggestions for robust-resistant smoothing involving a constrained form of smelting seem promising. (See "The use of smelting in guiding the re-expression" in this volume, for smelting in its simplest forms.)

These differences from one time series to another are as truly differences in data *structure* as the difference between a $7 \times 7 \times 7$ factorial and a 7×7 Latin Square.

VI. BACK TO THE "STYLE" COORDINATES

How did the choice of background occur?

Mainly through the choice of criteria -- choices influenced by the tools available to find consequences of certain sorts of criteria.

* starting out *

The simplest background with which we can work is one that is narrow (i.e. highly specified) and involves functions, etc. that are relatively manageable by accessible mathematics.

There should be no surprise that we started — and are likely to start new areas -- with an overutopian background which has just these characteristics.

Such starts have offered a foundation on which many important concepts -- such as power -- first flowered. (Both nonparametric and robust-resistant backgrounds trace back to overutopianism.)

* the nonparametric eras *

Too often we forget that R. A. Fisher (in *The Design of Experiments*) gave a crucial impetus to the nonparametric movement.

The seeds of nonparametric's decay were inherent in its formulation however -- perhaps inevitably so. One could -- as was easily shown by example--ask for validity in the face of a nearly unspecified situation. But we failed to learn how--or even whether it is possible—to ask for stringency in similar generality.

* robust-resistant DATs *

At the base of robust-resistant techniques lies the idea of "reasonable" situations, an idea we are slowly formulating more and more clearly. Some such idea seems likely to be essential. We want stringency, and we want it reasonably broadly--not just in some overutopian situation. It seems likely -- nay, almost certain -- that we cannot have it in all situations (at least for realistic sample sizes).

Robust/resistant techniques are developed with two different sorts of guidance. The simulators--often via Monte Carlo rather than simple empirical experimental sampling -- study the performance of estimates and tests (often best thought of as interval estimates):

- in (very) finite samples
- for few, carefully selected, situations.

Their results apply to real sample sizes and nearly real situations, but there remains a logical possibility of gaps in coverage.

The asymptotic theorists look to large neighborhoods and to the worst absolute performance therein. In effect, they change the definition of stringency from

$$stringency = \frac{\textit{how well you do}}{\left(\begin{array}{l}\textit{the best that can done}\\ \textit{in the given situation}\end{array}\right)}$$

to something nearer

$$stringency^* = \frac{\left(\begin{array}{l}\textit{how well you are sure to}\\ \textit{do throughout a close neighborhood}\end{array}\right)}{\left(\begin{array}{l}\textit{the best that one can be}\\ \\ \textit{sure do throughout a close neighborhood}\end{array}\right)}$$

Like the simulator's careful choice of situations in which to seek stringency, this sort of change keeps us away from being unduly influenced by "the Lorelei" -- by situations whose idiosyncrasies offer large amounts of extra, untrustworthy information. "Untrustworthy" because very nearby situations do not provide it.

In one way or another, our robust/resistant background must reflect Charlie Winsor's principle:

"All actual distributions are Gaussian in the MIDDLE!"

Now that the assumption of *un*stretched tails has lost its charm, and Gaussianity's heavy emphasis of "information" from the extreme tails doesn't mislead us anymore, central peakedness -- as in sampling from Cauchy or double exponential distributions -- seems to be the next most seductive source of untrustworthy information.

VII. THE COMPONENTS OF DATA ANALYSIS PROCESSES

A classic remark, which should never lose its relevance, is that data analysis is "like using an Erector or Meccano set, where we put together as many pieces as may be necessary for our present problem". We should be prepared to use this idea early and often.

But our attention in this section is to another sort of breakdown -- one clearly essential to exploratory analyses, and very helpful for critical or confirmatory ones.

* some kinds of components *

We can gain considerable insight by dividing up our processes of data analysis into "components". Today I would emphasize these kinds:

- ADES = **a**utomatic **d**ata **e**xpanders.
  (these make fewer numbers into more, in a reversible way, providing the raw material for OCONs and DDAPs)
- OCONs = **o**ptimistic **con**centrators
  (these make more into still fewer, in an irreversible way, providing summaries for use by OUTs and CDAPs)
- DDAPS = **d**iagnostic **d**ata **a**nalysis **p**rocedures
  (these provide humans [tomorrow, also automatic judgment processes] with the basis for deciding (a) what, still undone, needs to be done, (b) what to do next, (c) when to stop "flexing")
- OUTs = **out**put processes
  (these take the preoutput from OCONs or CDAPs and, often further condensing it, make output.)
- CDAPS = **c**ritical **d**ata **a**nalysis **p**rocedures
  (these take the preoutput from one or more OCONs, and provide standard errors, tests of directionality, confidence intervals (individual or simultaneous), etc.)
- SDAPs = **s**elective or stepwise **d**ata **a**nalysis procedures
  (these are OUTs that *selectively* output that part of what is available to them as preoutput that they automatically judge to be most interesting.)

(For some further discussion along these lines, see Tukey 1980.)

VIII. A SUBSTANTIAL EXAMPLE

Suppose that we have 10 measurements on each of 7 quantities -- and that it is sensible to try to compare these quantities in terms of these measurements.

The first thing it is natural to do is exemplified by taking each set of 10 numbers and replacing it by either 12 or 22 numbers in at least one of these three ways:

- mean, s^2, and 10 residuals
- mean, s^2, and 10 originals
- mean, s^2, 10 originals, and 10 residuals

In so doing we will have applied one of three ADEs (**a**utomatic **d**ata **e**xpanders) to all seven of the groups of 10.

A conventional statistics text is likely to focus (a, lightly) on the middle one of the 3 ADE's above and (b, strongly) on forgetting all except the 7 means and the 7 s^2 (or perhaps only the seven means and the mean, s^2, of the seven s^2). Either form of (b) is an example of an OCON (of an **o**ptimistic **con**centrator.

Indeed the conventional statistics book is likely to go on to (c) in the form of (c1) an F-test (or a q-test using the Studentized range), (c2) some t-tests, (c3) an F-test whose positive result will call forth t-tests, or (c4) some other multiple-comparisons procedure. Any of these would be an instance of a CDAP (of a **c**ritical/confirmatory **d**ata **a**nalysis **p**rocedure).

Rather more alert analysts might take the 7 $\bar{y}$'s and 7 s^2's, and plot ln s^2 vs. $\bar{y}$ (if there are $\bar{y}$'s ≤ 0) or, better when feasible, vs. ln $\bar{y}$. This would be an instance of a DDAP (of a **d**iagnostic **d**ata **a**nalysis **p**rocedure). If any apparent dependence of ln s^2 on ln $\bar{y}$ was recognized, the informed analyst would be likely to plan to replace y by y^p, for a value of p indicated by the slope of the plot, and restart the analysis. This would be an instance of a choice based upon human judgment.

* more about choosing *

Why do we think anyone would have chosen one of the ADEs mentioned above to start the analysis? It is easy to suggest one moderate good reason and one moderately poor one. We might be dealing with a subject-matter field -- like insurance -- where *totals* have overriding importance. Since means are interconvertible with totals when counts are known, this is a moderately good reason.

Indeed, if we really believed that the distribution of errors and fluctuations involved in our data was exactly Gaussian, then we would believe in the applicability of the mathematical results that show $\bar{y}$ and s^2 to be not only optimum summaries but, if the distributions were all exactly Gaussian, exhaustive of all information in each set of 10.

This last remark, when applicable (if ever), would assign the ADE's and OCON's -- as well as CDAPs -- to an *overutopian* statistical background.

A semi-careful analyst might try a different sort of DDAP on our 7 tests of 10 numbers, asking if they seem to behave the way 7 samples of 10 from Gaussian distributions would.

If such a DDAP says "no!", such an analyst is likely to want to repeat the analysis with a different ADE, perhaps replacing each batch of 10 raw values by:

- the value of their biweight and the value of their s_{bi}^2 (Mosteller and Tukey 1977, Chapter 10), as well as 10/20 originals and/or residuals.

We now know (Kafadar 1980) that a CDAP that puts biweight and s_{bi}^2 into t-tests and simple multiple comparison procedures is valid, provided only we (i) have groups of close to 10 or more and we (ii) reduce the usual values for "degrees of freedom" by multiplication by 0.8.

A truly careful analyst would not have trusted a conventional DDAP to detect lack of Gaussianity. Instead he/she would run *both* a classical ADE *and* a robust/resistant ADE at the beginning, planning to continue with OCON's and CDAP's appropriately, and to compare the results of the two chains of components as an informal DDAP — one where substantial discrepancy would call for much more detailed diagnosis.

Such an analyst would be starting with *two* ADE-OCON-CDAP chains, one whose stochastic background is *overutopian* and whose broad stringency is *dubious*, and one of *high broad stringency* whose background is *robust/resistant*.

Besides ADEs that focus on mean -- or biweights -- we can equally well have ADEs for single batches that focus on

medians and hinge-spreads (interquartile separations)

- trimmed means (means omitting some values of each end -- $n/4$ for midmeans, $(3+n)/5$ for abmeans).
- other estimates of center
- all above

The point here is there can be many ADEs for what may seem to be the same data structure.

If our 7×10 body of data had had 2-way structure, the simplest ADEs would make it into a $(7+1) \times (10+1)$ structure with 88 numbers (26% more than 70).

If we had had 128 values in a 2^7, the simplest ADE would convert this to a $(2+1)^7$ involving 2187 numbers -- many pairs of which would of course be either the same or the same except for sign.

Some "data expanders" really expand!

Looking more broadly at this example, the sequence starting

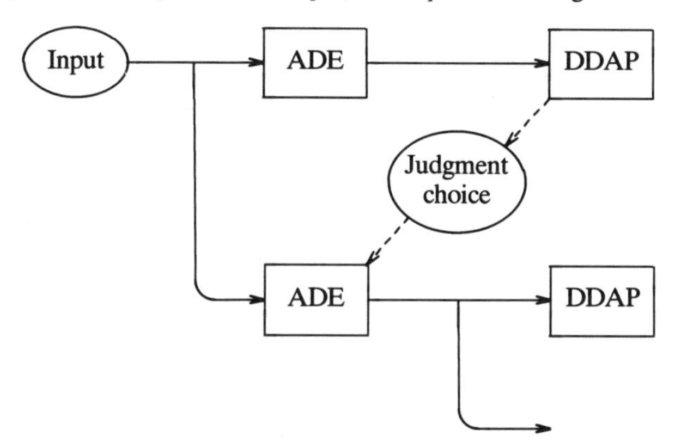

is typical of *exploratory* data analysis while the once-through sequence

in ——→ ADE ——→ OCON ——→ CDAP ——→ out

is typical of (the naive picture of) *critical/confirmatory* data analysis.

The earlier sections discuss the classifications that underlie the labels we have used in connection with this example.

REFERENCES

Cleveland, W. S., Dunn, D. M., and Terpenning, I. J. (1979). SABL -- a resistant seasonal adjustment procedure with graphical methods for interpretation and diagnosis, in *Seasonal Analysis of Economic Time Series,* ed. A. Zellner, U.S. Dept. of Commerce, Bureau of the Census.

Kafadar, K. (1980). Using biweights in the 2-sample problem. Princeton Technical Report #152, Princeton University.

Mallows, C. L., and Tukey, J. W. (1981). An overview of techniques of data analysis, emphasizing its exploratory aspects, in *Some Recent Advances in Statistics,* forthcoming from Academic Press.

Mosteller, F. and Tukey, J. W. (1977). *Data Analysis and Regression,* Addison-Wesley, Reading, Mass.

Pitman, E. J. G. (1937). Significance tests which may be applied to samples from any population. *Suppl. J.R. Statist. Soc.,* **4,** 119, 225.

Tukey, J. W. (1977). *Exploratory Data Analysis,* Addison-Wesley, Reading, Mass.

Tukey, J. W. (1980). Styles of data analysis and their implications for statistical computing, in *COMPSTAT 1980: Proceedings in computational statistics,* eds. M. M. Barritt and D. Wishart, Physica-Verlag.

SOME MULTIPLE Q-Q PLOTTING PROCEDURES

Nicholas P. Jewell[1]

Department of Statistics
Princeton University

I. Q-Q PLOTS

This paper is concerned with some extensions to the idea of a quantile-quantile (Q-Q) plot which is commonly used by a statistician for a variety of purposes. We begin by reminding ourselves exactly what a Q-Q plot is. An excellent description of Q-Q plots along with several examples can be found in Wilk and Gnanadesikan (1968). A description is also given in Gnanadesikan (1977).

Suppose we have two distribution functions F_1 and F_2. The Q-Q plot is a graphical way of comparing F_1 and F_2 with regard to their shape. It is simply a plot of points whose x-coordinate is the pth quantile of F_1 and whose y-coordinate is the pth quantile of F_2, there being one point for each value of p between 0 and 1. In other words the <u>Q-Q plot of F_1 against F_2</u> is a plot of the points $\{(q_1(p), q_2(p))\}$ as p varies between 0 and 1. This is illustrated in Figure 1.

[1]Research supported in part by a contract with the Office of Naval Research, No. N00014-79-C-0322, awarded to the Department of Statistics, Princeton University, Princeton, N.J.

ISBN 0-12-438180-4

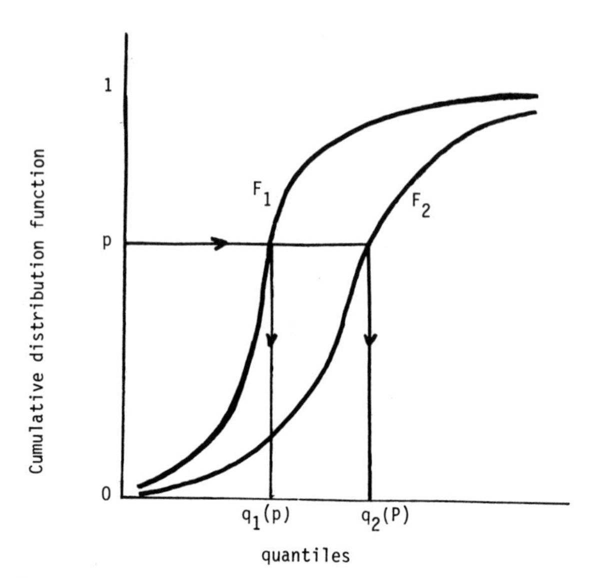

FIGURE 1. Illustration for Q-Q plot.

What will the Q-Q plot look like if F_1 is actually equal to F_2? It will be a straight line of slope 1 through the origin. What if F_1 differs from F_2 only through a scale and location change? It will still be a straight line with the slope and intercept depending on the values of the scale and location change. For future reference we note that if $F_2(x) = F_1((x-\mu)/\sigma)$ then the slope of the straight line will be σ and the intercept will be μ/σ. What if F_1 differs from F_2 in a more fundamental way? The Q-Q plot will no longer be a straight line. Information concerning the differences between F_1 and F_2 may be inferred from the shape of the Q-Q plot. We refer to Wilk and Gnanadesikan (1968) and Gnanadesikan (1977) for details. As an example, Figure 2 illustrates the Q-Q plot of the uniform distribution

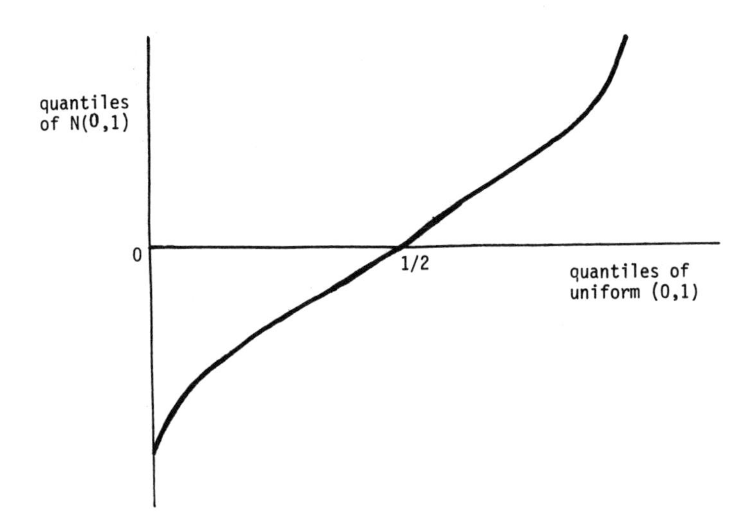

FIGURE 2. Q-Q plot of uniform distribution on (0,1) against the standard Gaussian distribution.

on (0,1), i.e. U(0,1), against the standard Gaussian distribution, i.e., N(0,1).

Figure 1 assumes that F_1 and F_2 are given by smooth curves, but this is merely for illustrative convenience. The definition of a Q-Q plot that we have given applies equally well, for example, to the case when either F_1 or F_2 or both are step functions.

A common use of the Q-Q plot in comparing the shape of two distribution functions F_1, F_2 is when one of the distribution functions, say F_1, is the empirical cumulative distribution function (ecdf) generated by a set of independent observations following an unknown distribution law. The other, F_2, is a theoretical distribution function whose law the observations are suspected to follow. The quantiles of the ecdf generated by a set of n independent observations $x_1,\ldots,x_n$ are given by

$$q(p) = \begin{cases} x_{(j)} & \text{for } (j-1)/n<p\leq j/n \quad (0<j\leq n-1) \\ x_{(n)} & \text{for } p = 1 \end{cases}$$

where $x_{(1)},\ldots,x_{(n)}$ represent the order statistics for the n observations. As the quantiles of the ecdf do not change as p varies from $(j-1)/n$ to j/n it is customary to plot the quantiles only for the values $p = 1/n, 2/n,\ldots,1$. Thus, in this case, the Q-Q plot consists of a plot of the points $(\tilde{x}_j, x_{(j)})$ for $j = 1,\ldots,n$ where $\tilde{x}_j$ is the quantile of F_2 corresponding to the cumulative probability $p = j/n$. In practice another adjustment is made to cope with the case $p=1$ since for many continuous distributions the quantile corresponding to $p = 1$ is infinite. Losing only one point from the plot may not seem important but this point occurs in the tails of the distributions which is often where we would like as much information as possible. One possible adjustment is to take $\tilde{x}_j$ as the quantile of F_2 corresponding to a cumulative probability $p_j = (j-1/3)/(n + 1/3)$. Figure 3 illustrates a Q-Q plot where F_1 is the ecdf generated by 100 independent observations from $U(0,1)$ and F_2 is $N(0,1)$.

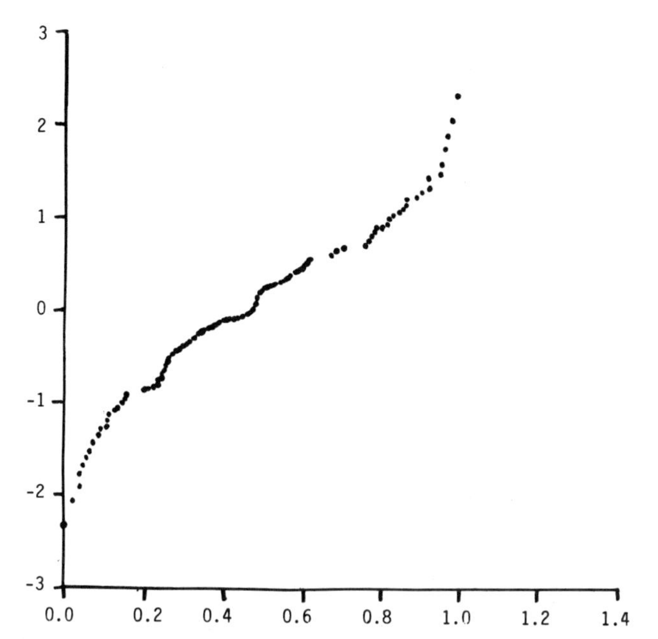

FIGURE 3. Q-Q plot of ecdf generated by 100 U(0,1) random variates against N(0,1).

Purposes of a Q-Q Plot

We summarize some of the common purposes of a Q-Q plot:

(i) check the distributional properties of a set of observations against a prespecified 'model' distribution;

(ii) obtain information about the 'true' distribution governing the observations in comparison with some standard reference distribution, e.g. information about symmetry, kurtosis, heavy or light tails, etc..

Some common practical uses of the Q-Q plot are

(i) check for normality (not always on original data points but sometimes on residuals after a model has been fitted);

(ii) check on hypothesized survival, reliability or extreme value distributions, e.g. exponential, Weibull, Gumbel distributions; (adjustments to the basic definition of a Q-Q plot of an ecdf against a hypothesized distribution can be made in order to cope with certain kinds of censored data which often arises in these areas of application).

We remark here that some people may be more familiar with Q-Q plots of an ecdf against N(0,1) under the name of plotting data on Normal probability graph paper.

II. THE MULTIPLE Q-Q PLOT FOR EXTREME VALUE DATA

It sometimes occurs that a set of observed data points are each believed to represent a maximum of an unobserved set of data points. For example, in a set of maximum monthly rainfalls each point may represent a maximum of about 30 observations on daily rainfalls. Other examples are maximum daily temperatures or greatest yearly floods. Often the method in which the data is collected infers that the maxima are maxima of a very large unobserved set of data points (essentially an infinite number of points if readings are taken continuously but only maxima recorded). It is of great interest to model this data in some fashion in order to calculate the chances of a very high value occurring within a certain period of future time or reciprocally the expected amount of time to the first occurrence of a certain extreme value. For example, in construction of a tall building the chances of a 100 m.p.h. wind speed occurring within the building's projected lifetime may be important in decision-making on construction details.

The theory underlying model building for this type of extreme value data relies on the following well-known observation and theorem.

Observation. The largest element, $x_{(m)}$, of a set of m independent observations, $x_1,\ldots,x_m$, drawn from a distribution F follows the distribution F^m.

Theorem. If the sequence of distribution functions F^m converge in law to a distribution function (after appropriate scaling and translation) as $m \to \infty$, then this asymptotic limit distribution can be one of only 3 distributions:

$$\text{(i)} \quad F_1(x) = \begin{cases} \exp(-x^{-\alpha}) & x>0 \quad \alpha>0 \\ 0 & x\leq 0 \end{cases}$$

$$\text{(ii)} \quad F_2(x) = \begin{cases} \exp(-(-x)^{\alpha}) & x\leq 0 \quad \alpha>0 \\ 1 & x>0 \end{cases}$$

$$\text{(iii)} \quad F_3(x) = \exp(-e^{-x}) \qquad -\infty<x<\infty$$

We refer to the book by Galambos (1978) for a detailed discussion of the background of this theorem and several different extensions.

The distribution F_3 is known as the standard Gumbel distribution and we shall denote it by $G(0,1)$. The Gumbel distribution with location parameter a and scale parameter b is given by $F_3(x) = \exp(-\exp(-(x-a)/b))$. The Gumbel distribution is the one of the three asymptotic distributions which is most often used when fitting extreme value data. Also simple transformations reduce F_1 and F_2 to F_3.

As a consequence of the observation and theorem we often hope that our extreme value data follows, at least approximately, one of the 3 extreme value distributions, say F_3. We then estimate location and scale parameters using, for example, the method of moments or maximum likelihood. We can graphically check and examine our distributional assumptions by using a Q-Q plot of the ecdf against G(0,1). We refer to Gumbel (1958) for a general description of the statistics of extreme value data.

The purpose of this paper is to suggest that we can squeeze a little more information out of Q-Q plots about the question at hand, namely how 'close' is the 'true' distribution of our data points to a Gumbel distribution. For example it might occur to us that we would be better using maximum daily rainfalls where the maximum is taken over 1 year rather than 12 separate maximum monthly rainfalls. We could then try fitting a Gumbel distribution to these 'new' data points. This would reduce the amount of data for estimation purposes but this reduction might be advisable if our original observations follow a distribution which is quite far from a Gumbel distribution leading to inaccurate parameter estimates. At any rate it would be useful to know whether taking maxima over a larger set of underlying observations might take us significantly closer to a Gumbel distribution and whether there is any information about if the distribution of interest is in fact converging to a Gumbel distribution as we take maxima over successively larger sets of independent observations. A diagnostic method of examining these questions is the purpose of the multiple Q-Q plot.

The multiple Q-Q plot for extreme-value data is based on the simple observation that the maximum of m independent observations drawn from a Gumbel distribution with location parameter a and scale parameter b (denoted by G(a,b)) follows the Gumbel distribution $G(a+b\log_e m, b)$. Hence the Q-Q plot of the maxima of sets of m independent and identically distributed observations from G(a,b) against G(0,1) is a straight line with slope b and intercept $a+b\log_e m$. If we draw these plots for different m on the same graph we have a set of parallel lines each shifted a certain amount from the preceding one. If we simply draw the plots corresponding to m=1,2,4,8,..., we have a set of parallel lines where the intercept is increased the same amount, $b\log_e 2$, each time.

Before considering the multiple Q-Q plot calculated from

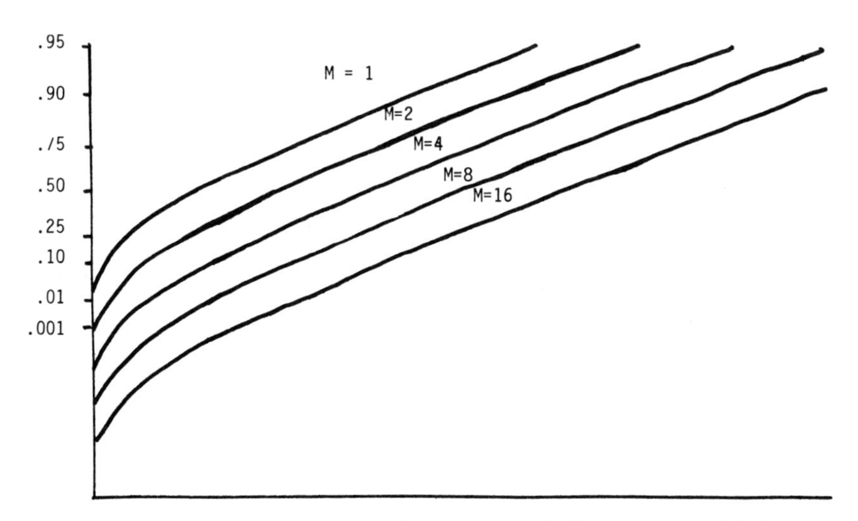

FIGURE 4. The multiple Q-Q plot of the exponential distribution against G(0,1).

a set of data points we will look at some theoretical multiple Q-Q plots of familiar distributions against G(0,1). Figure 4 illustrates the Q-Q plot of the exponential distribution (M=1) and the maximum of M independent and identical exponentials (for M=2,4,8,16) all against G(0,1). The scale on the y-axis is given in terms of the cumulative probability p rather than the values of the quantiles q(p). Figures 5 and 6 illustrate the same graphs for the logistic distribution and the Gaussian distribution respectively.

These pictures show us

(i) where each of the distributions differ from the Gumbel, and

(ii) convergence is slow (the Q-Q plots for M=16 don't appear any straighter than those for M=1) although, for M=16, the differences from the Gumbel can be seen to occur further

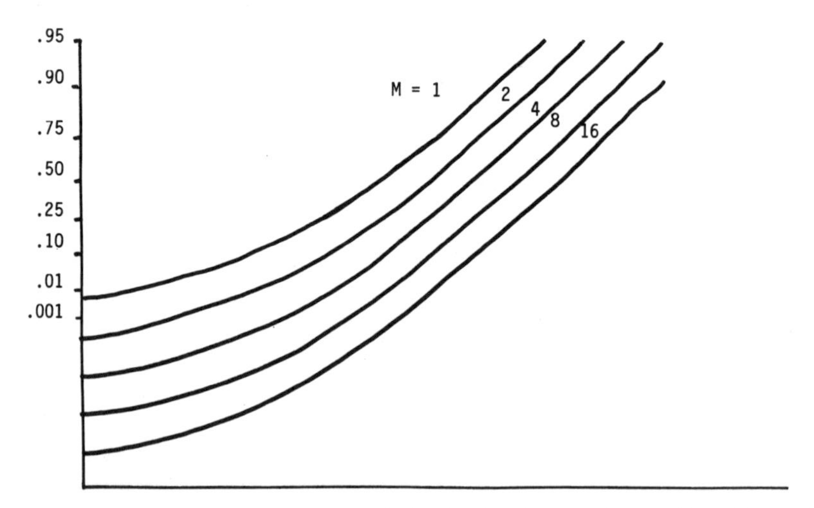

FIGURE 5. The multiple Q-Q plot of the logistic distribution against G(0,1).

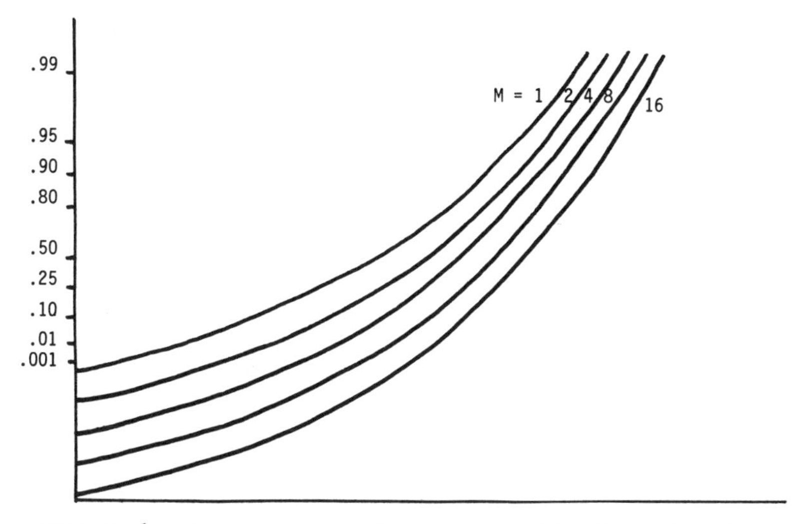

FIGURE 6. The multiple Q-Q plot of the Gaussian distribution against G(0,1).

out in the tail. For example, in Figure 4, for M=1 the Q-Q plot is straight only from p=0.5 to 1 whereas for M=16 the plot is straight from p=0.1 to 1.

It seems reasonable to infer from these graphs (if we didn't know about our theorem) that the maxima of exponential, logistic and normal distributions may be converging to a Gumbel distribution although the convergence is slow.

Can we obtain the same sort of information from a finite set of data points drawn from an exponential distribution, say, rather than from knowledge of the complete distribution? The multiple Q-Q plot for a set of data points attempts to achieve this.

Suppose we have a set , $\{x_1,\ldots,x_n\}$,of data points. For the multiple Q-Q plot all possible pairs of points, (x_i,x_j) for $i\neq j$, from the original data set are examined and a 'new' data set is created consisting of the maxima of these pairs. $F_{n(2)}$ is the ecdf of these $\binom{n}{2}$ points. Of course the points in this set are just elements of the original data set each point occurring with the following frequencies:

$x_{(1)}$ does not appear,

$x_{(j)}$ appears $\binom{j-1}{1}$ times for $2 \leq j \leq n$.

Hence $F_{n(2)}$ is a step function with jumps occurring only at $x_{(2)}, \ldots, x_{(n)}$ and its values at these points are given by

$$F_{n(2)}(x_{(j)}) = \sum_{i=2}^{j} \binom{i-1}{1} / \binom{n}{2} = \binom{j}{2} / \binom{n}{2} \qquad (j \geq 2).$$

Of course the new data points are not independent since different pairs may contain the same original point. However, since

$$F_{n(2)}(x) = \left[1/\binom{n}{2}\right] \sum_{\text{all pairs}} I_{\{\text{max. of pair}(x_i, x_j) \leq x\}}$$

(where I_A is the indicator function which is 1 or 0 according to whether the statement A is true or false respectively),

we have $E(F_{n(2)}(x)) = F^2(x)$ where F is the distribution governing the original data set. That is, for each x, $F_{n(2)}(x)$ is an unbiased estimate of $F^2(x)$. Thus, just as we expect the Q-Q plot of the original ecdf F_n against a pre-specified distribution G to reflect the similarities and differences between F and G so might we reasonably expect the Q-Q plot of $F_{n(2)}$ against G to illustrate the similarities and differences between F^2 and G.

In a similar fashion, for $m \leq n$, we can take all possible

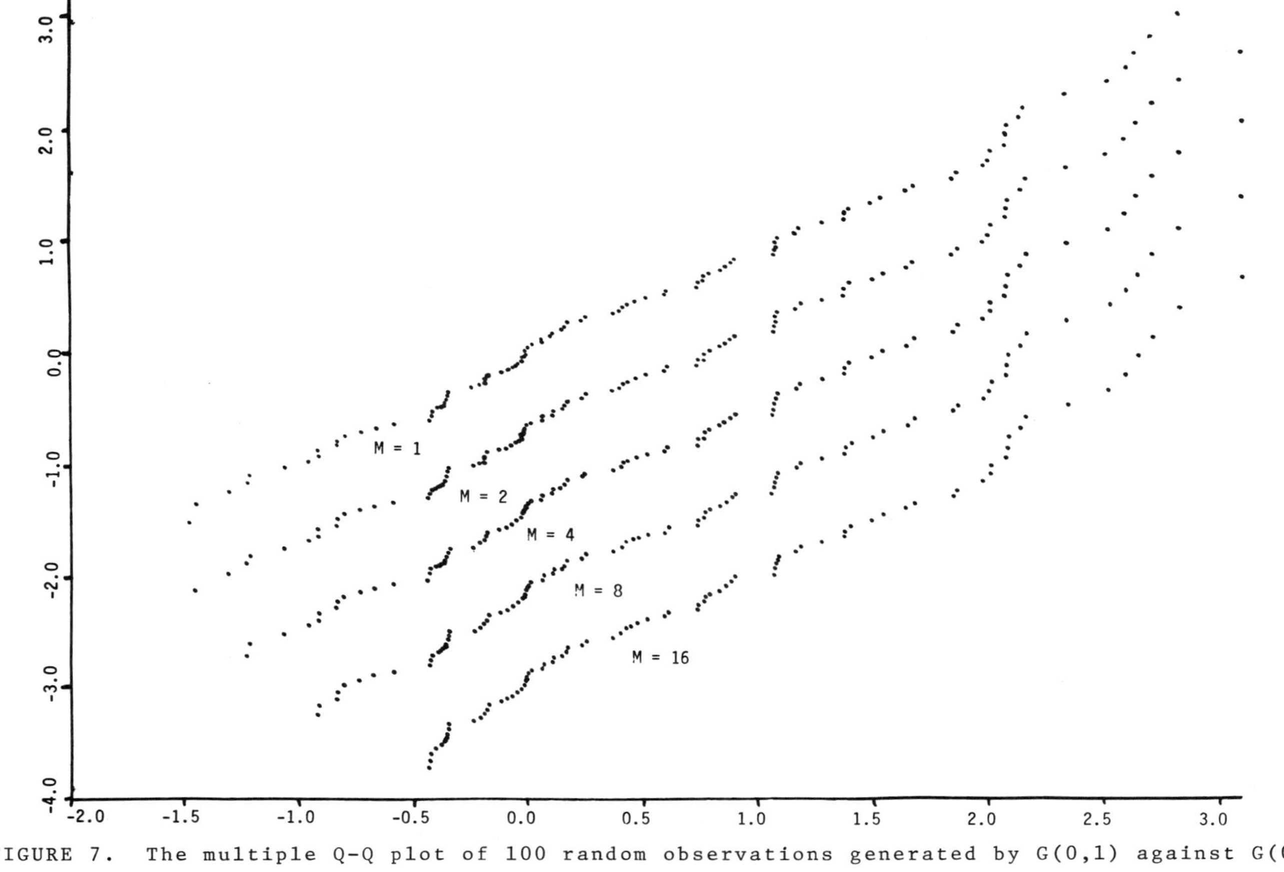

FIGURE 7. The multiple Q-Q plot of 100 random observations generated by G(0,1) against G(0,1).

different m-tuples containing m distinct x_j's and create a new data set consisting of the maxima of these $\binom{n}{m}$ m-tuples. We denote the ecdf of these points by $F_{n(m)}$. Again it is easy to see that $F_{n(m)}$ is a step function with jumps occurring only at $x_{(m)}, x_{(m+1)}, \ldots, x_{(n)}$ and its values at these points are given by

$$F_{n(m)}(x_{(j)}) = \sum_{i=m}^{j} \binom{i-1}{m-1} / \binom{n}{m} = \binom{j}{m} \binom{n}{m} \qquad (j \geq m)$$

Furthermore $E(F_{n(m)}(x)) = F^m(x)$ for each x.

The multiple Q-Q plot against G(0,1) is then a plot on the same graph of the Q-Q plots of F_n against G(0,1) and $F_{n(m)}$ against G(0,1) for m = 2,4,8,...etc.. If the original data arises from a Gumbel distribution G(a,b), we might expect that these Q-Q plots would be parallel straight line plots, each shifted in intercept by $b\log_e 2$ from the preceding plot.

We will now look at some examples. Figure 7 is the multiple Q-Q plot against G(0,1) for a set of 100 random data points generated from G(0,1). In one sense this multiple Q-Q plot illustrates the best behavior we can hope for. Figures 8, 9, and 10 are the multiple Q-Q plots against G(0,1) for data sets of 200 random data points generated from the standard Gaussian, logistic and exponential distributions respectively. It is interesting to compare these figures with Figures 4, 5, and 6.

Lack of space precludes looking at further examples but already these figures illustrate that we can obtain diagnostic information from multiple Q-Q plots of a single sample of

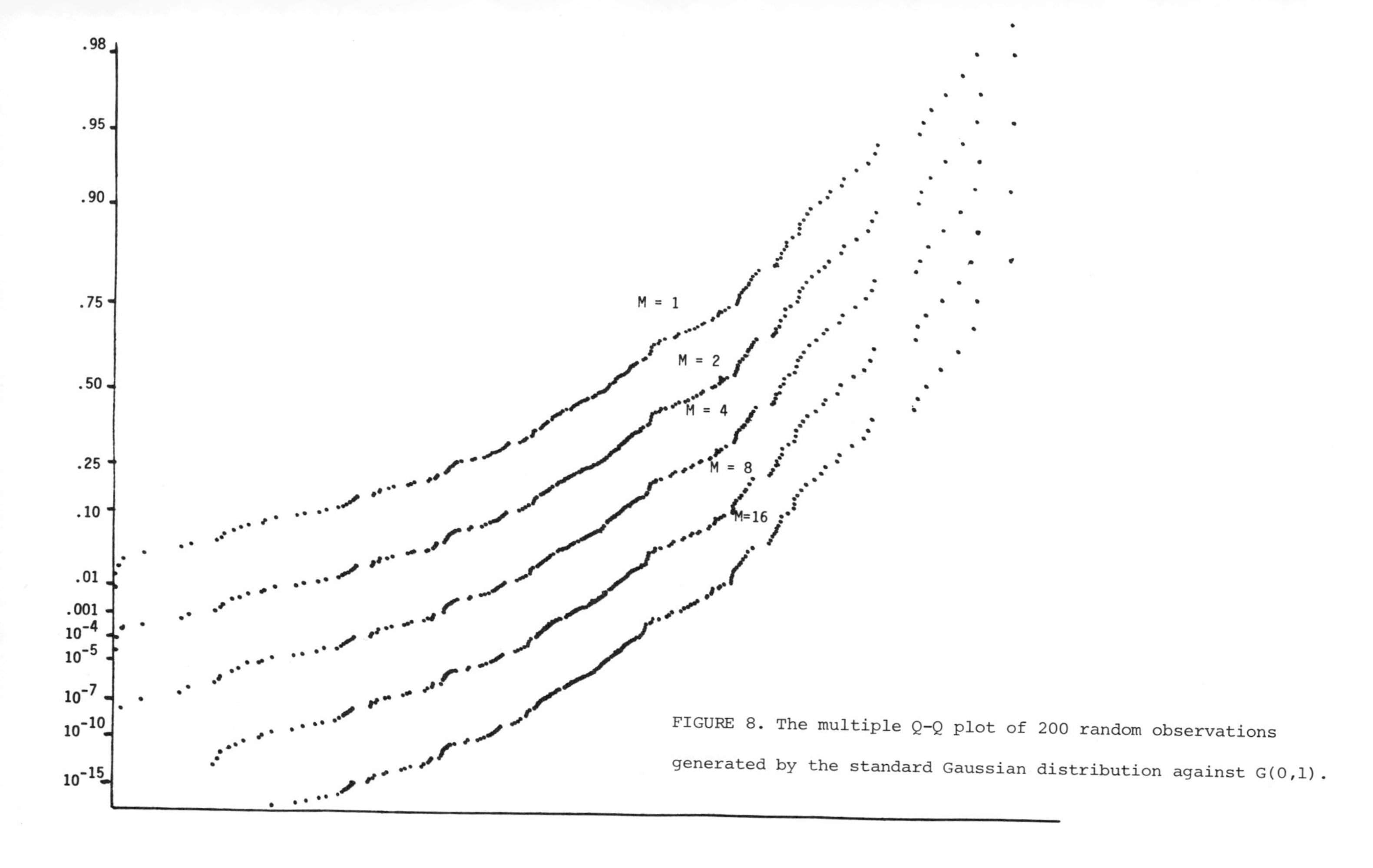

FIGURE 8. The multiple Q-Q plot of 200 random observations generated by the standard Gaussian distribution against G(0,1).

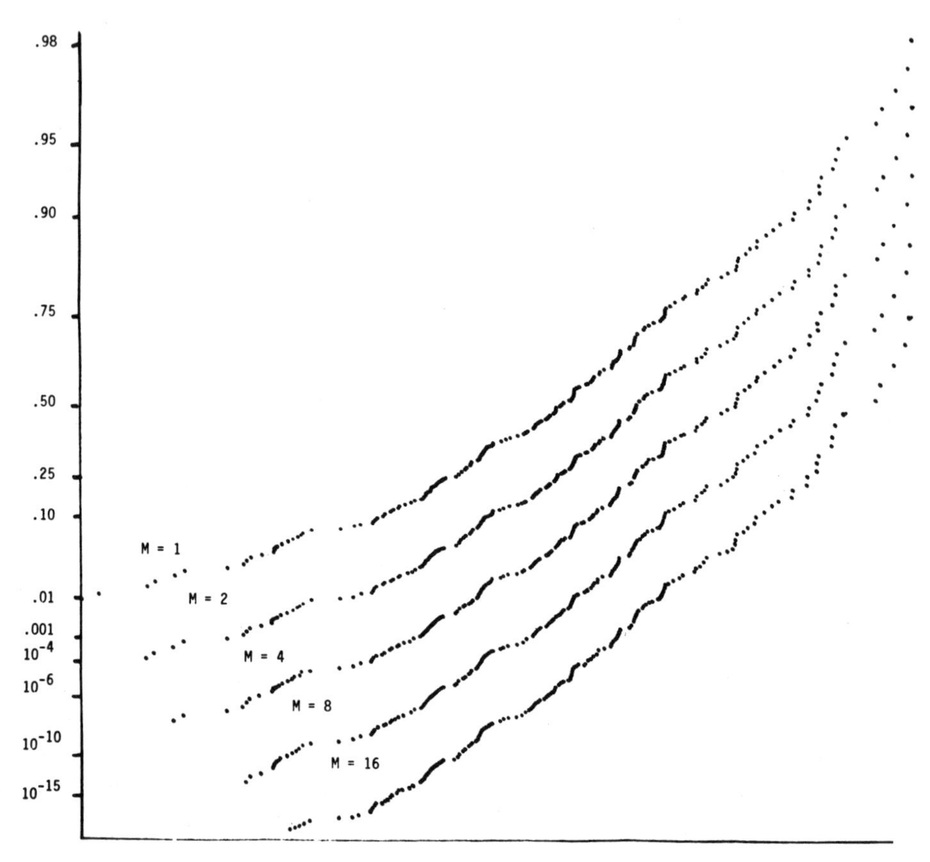

FIGURE 9. The multiple Q-Q plot of 200 random observations generated by the logistic distribution against G(0,1).

observations relating to how close the observed distribution is to G(0,1) and also some qualitative information about convergence of maxima and the rate of convergence. It would be satisfying to see a multiple Q-Q plot of lines which 'straighten' out as m becomes larger suggesting that taking maxima of a larger number of underlying measurements would achieve a closer fit to a Gumbel distribution. The fact that this does not occur throughout the complete range for the

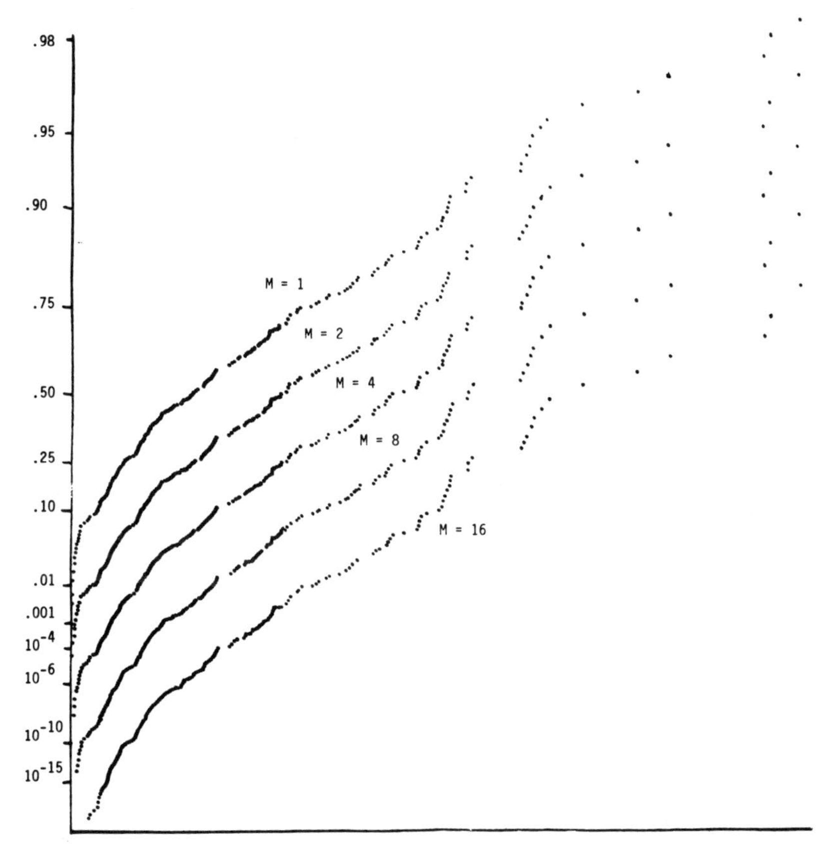

FIGURE 10. The multiple Q-Q plot of 200 random observations generated by the exponential distribution against G(0,1).

illustrated distributions (although it does occur in a restricted range) is due to the notoriously slow convergence to the extreme value distribution. Nevertheless the multiple Q-Q plot has provided us with some additional diagnostic information that the single Q-Q plot did not.

How Far Should We Go?

Clearly we cannot continue taking maxima of larger and larger subsamples and then draw meaningful Q-Q plots; e.g. if m=n then we only have one point for the Q-Q plot! A preliminary look at some theoretical results suggests that we can allow m to be as large as $n^{1/2}$ and still draw sensible Q-Q plots. The theoretical details will appear in a future paper by the author.

III. COMMENTS AND OTHER MULTIPLE Q-Q PLOTS

1. It is simple to adjust a program written to draw Q-Q plots to enable it to provide multiple Q-Q plots for extreme value data.

2. Two extensions to our multiple Q-Q plot may be very useful. The first is to adjust the definition of the Q-Q plot in order to make the plot a horizontal line if the shapes of the distributions being compared are the same. It may be easier for the eye to check whether several lines are parallel if they lie on the horizontal. Second it may be interesting to compare the data at the 2^k - tuple stage with that at the 2^{k+1} - tuple stage. This might provide information about how much change in the distribution of the maxima was occurring by taking maxima of 2^{k+1} points rather than 2^k. If the distribution of the maxima of 2^k points is close to a Gumbel distribution then so will be the distribution of the

maxima of 2^{k+1} points. Thus the Q-Q plot comparing these two subsample distributions should be close to a straight line.

3. Multiple Q-Q plots are not confined to the analysis of extreme-value data. In fact some earlier comments suggest that we might look at situations where the distributional convergence of interest is of a faster rate than that for extreme order statistics. The most obvious case is the convergence of certain sample averages to a Gaussian distribution.

The Multiple Q-Q Plot for Sample Averages

As noted in the introduction one of the most common usages of the Q-Q plot is to compare an ecdf with the standard Gaussian distribution. Often the investigator is not only concerned with the distribution of the original data but might also like information on whether sample averages have distributions which tend to look like the Gaussian as the sample size increases. It is this sort of information which the multiple Q-Q plot for sample averages attempts to provide.

The multiple Q-Q plot for sample averages will be described in detail in a future paper by the author. We will briefly describe it here and look at some illustrations. As for the extreme value data multiple Q-Q plots we look at subsamples of the original data set. Instead of taking maxima we now construct weighted averages: in fact $m^{1/2}$ times the subsample average where m is the size of the subsample. The main difference from the extreme value multiple Q-Q plot is that the 'new' data set constructed in this way consists of points which are not members of the original data set. In fact there will be $\binom{n}{m}$ 'new' points, in general, which is a large number to look at if m is small. There are several ways round this -- details will appear elsewhere.

Figures 11, 12 and 13 illustrate one version of the multiple Q-Q plot for sample averages for 100 random observations generated by N(0,1), U(0,1) and the Cauchy distribution

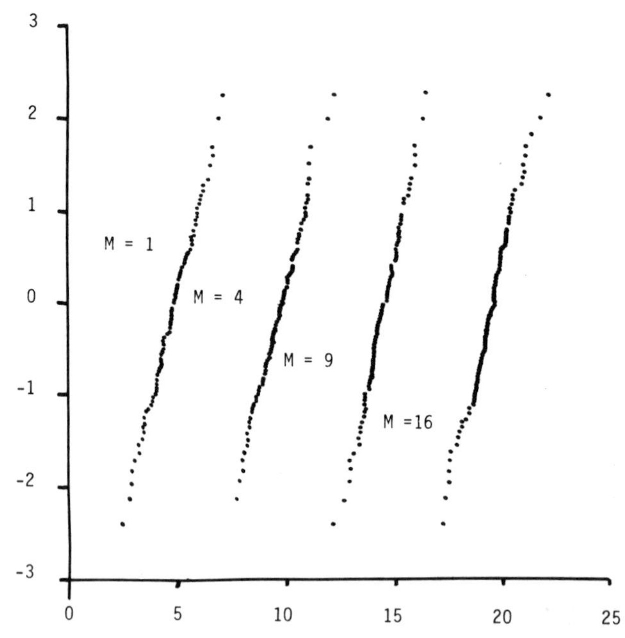

FIGURE 11. The multiple Q-Q plot for sample averages of 100 random observations generated by N(0,1) against N(0,1).

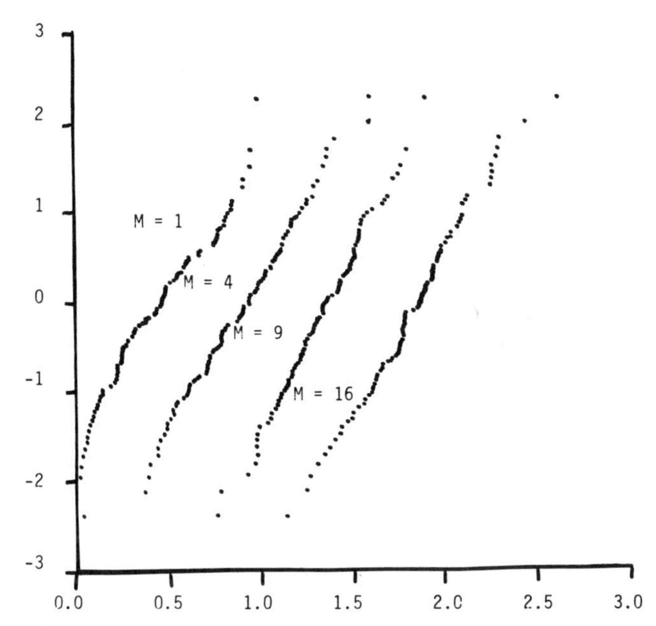

FIGURE 12. The multiple Q-Q plot for sample averages of 100 random observations generated by U(0,1) against N(0,1).

with location parameter 5.0 respectively, all against N(0,1). The subsamples chosen were m=1,4,9,16. We leave the figures to speak for themselves concerning the different types of behaviour which are possible.

The Future

The author is presently investigating ideas on how to use the techniques of empirical sampling applied here in order to make some standard hypotheses tests and estimation methods less susceptible to incorrect distributional assumptions. The idea is to 'robustify' tests and estimates by

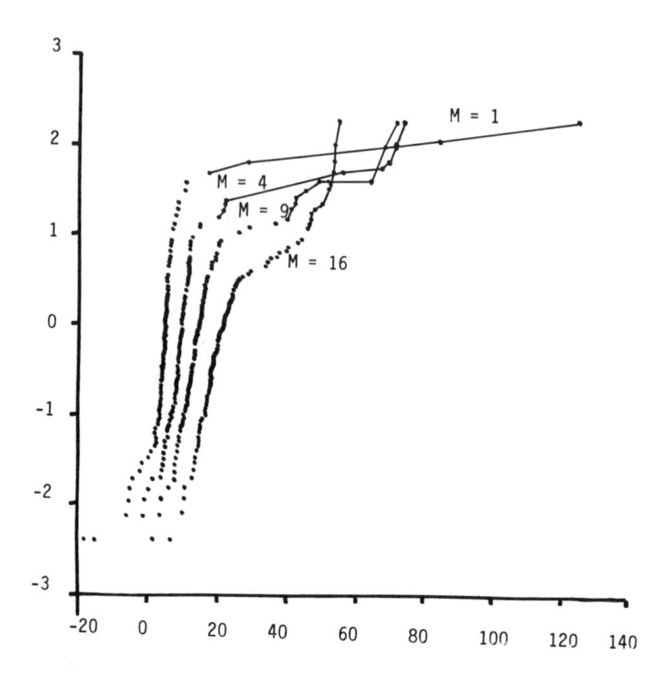

FIGURE 13. The multiple Q-Q plot for sample averages of 100 random observations generated by the Cauchy distribution with location parameter 5.0 against N(0,1).

exchanging 'what you are not sure about' for assumptions 'which you are sure about'.

ACKNOWLEDGMENTS

The author would like to thank P. Bloomfield and S. Zeger for several stimulating discussions on the topics of this paper.

REFERENCES

Galambos, J. (1978). "The Asymptotic Theory of Extreme Order Statistics," Wiley, New York.

Gnanadesikan, R. (1977). "Methods for Statistical Data Analysis of Multivariate Observations," Wiley, New York.

Gumbel, E.J. (1958). "Statistics of Extremes," Columbia University Press, New York.

Wilk, M.B., and Gnanadesikan, R. (1968). Probability plotting methods for the analysis of data. *Biometrika* 55, 1-17.

A Reader's Guide to Smoothing Scatterplots and Graphical Methods for Regression

William S. Cleveland

Bell Laboratories
Murray Hill, New Jersey 07974

1. Introduction

The major thrust of the talk at the conference was smoothing scatterplots and graphical methods for regression. Since the written detail of the second topic will shortly appear as a chapter in a book (Chambers et al., to appear) and the written detail of the first topic has already appeared (Cleveland, 1979), I will attempt only to review the major points and steer the reader to other literature.

2. Smoothing Scatterplots

Most of the plots that form the collection of regression graphics are scatterplots of one variable against another. While a scatterplot by itself is a powerful analytic tool, its visual message can be substantially increased by smoothing. Suppose (x_i, y_i), $i = 1,...,n$ are the points of the scatterplot. The dependence of Y on X can be summarized by another set of points, $(x_i, \hat{y}_i)$, where $\hat{y}_i$ portrays the location of the distribution of Y given $X = x_i$. The point $(x_i, \hat{y}_i)$ is called the smooth point at x_i and $\hat{y}_i$ is called the fitted value at x_i.

One example of the usefulness of smoothing scatterplots begins with Figure 1. Horowitz (to appear) in his very sensible article on the distribution of air pollution maxima, regressed the logarithms of daily maximum ozone concentrations on a quadratic polynomial of time; the data are from a site in the St. Louis metropolitan area and run for about one year beginning in January. Figure 1 is a plot of the residuals, r_t, against time, t. If the polynomial has done a proper job of describing the mean function of the log concentrations, then r_t should have a mean close to 0 for all t. But it is hard to make the judgment of the mean of r_t from the scatterplot alone. In Figure 2 the scatterplot and a set of smooth points, which are portrayed by joining successive points by straight lines, are shown. Now the judgment can be made with more certainty. The

ISBN 0-12-438180-4

plot suggests a small amount of wandering of the level of the series and therefore a slight lack of fit of the quadratic. My inclination would be to fit a trigonometric polynomial

$$\mu + \sum_{j=1}^{p} \alpha_j \sin[(2\pi tj)/365] + \beta_j \cos[(2\pi tj)/365]$$

to the log concentrations. Since a second harmonic appears to be emerging in the smooth points in Figure 2, p could be expected to be at least 2.

The method used to smooth the scatterplot in Figure 2 is robust locally weighted regression (Cleveland, 1979). One salient feature of this smoothing procedure is robustness, which means that a small fraction of outliers cannot distort the fitted values. While the example in Figures 1 and 2 is a time series, so that the abscissas are equally spaced, robust locally weighted regression can also be used to smooth general scatterplots with unequally spaced abscissas. Other methods in the literature for smoothing scatterplots are referenced in (Cleveland, 1979) and (Stone, 1977).

3. Graphical Methods for Regression

The major points in the talk for regression graphics are:

(1) Plotting the response against each explanatory variable is a dangerous practice.

(2) Graphical looks into the multidimensional design configuration (the n points in p-space formed by the rows of the design matrix) can be helpful.

(3) The adjusted variable plot for the k-th explanatory variable characterizes the estimation of the regression coefficient for that variable and can detect lack of fit. This display is made by plotting the residuals from regressing the response on all explanatory variables except the k-th against the residuals from regressing the k-th explanatory variable on all others, and smoothing the plot.

(4) Plotting residuals against fitted values and smoothing the plot can detect lack of fit.

(5) Plotting residuals against fitted values is not a good way to detect a dependence of the scale of the errors on the mean of the response. A better procedure is to plot the absolute residuals against the fitted values together with a smooth of the plot.

(6) If least squares is used the normality of the error terms should be checked by a normal probability plot of the residuals.

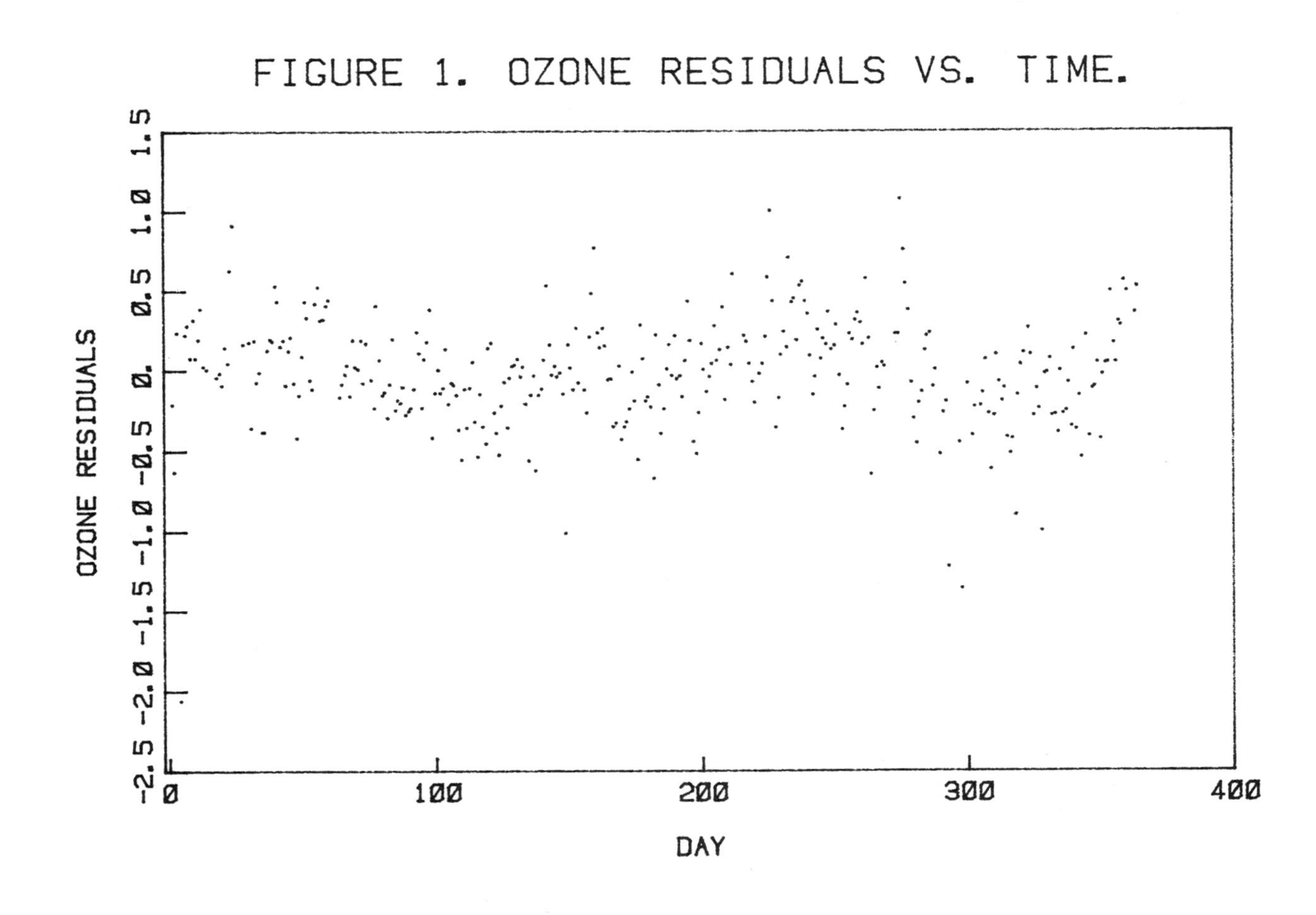
FIGURE 1. OZONE RESIDUALS VS. TIME.
OZONE RESIDUALS
1.5
1.0
0.5
0.
-0.5
-1.0
-1.5
-2.0
-2.5
0
100
200
300
400
DAY

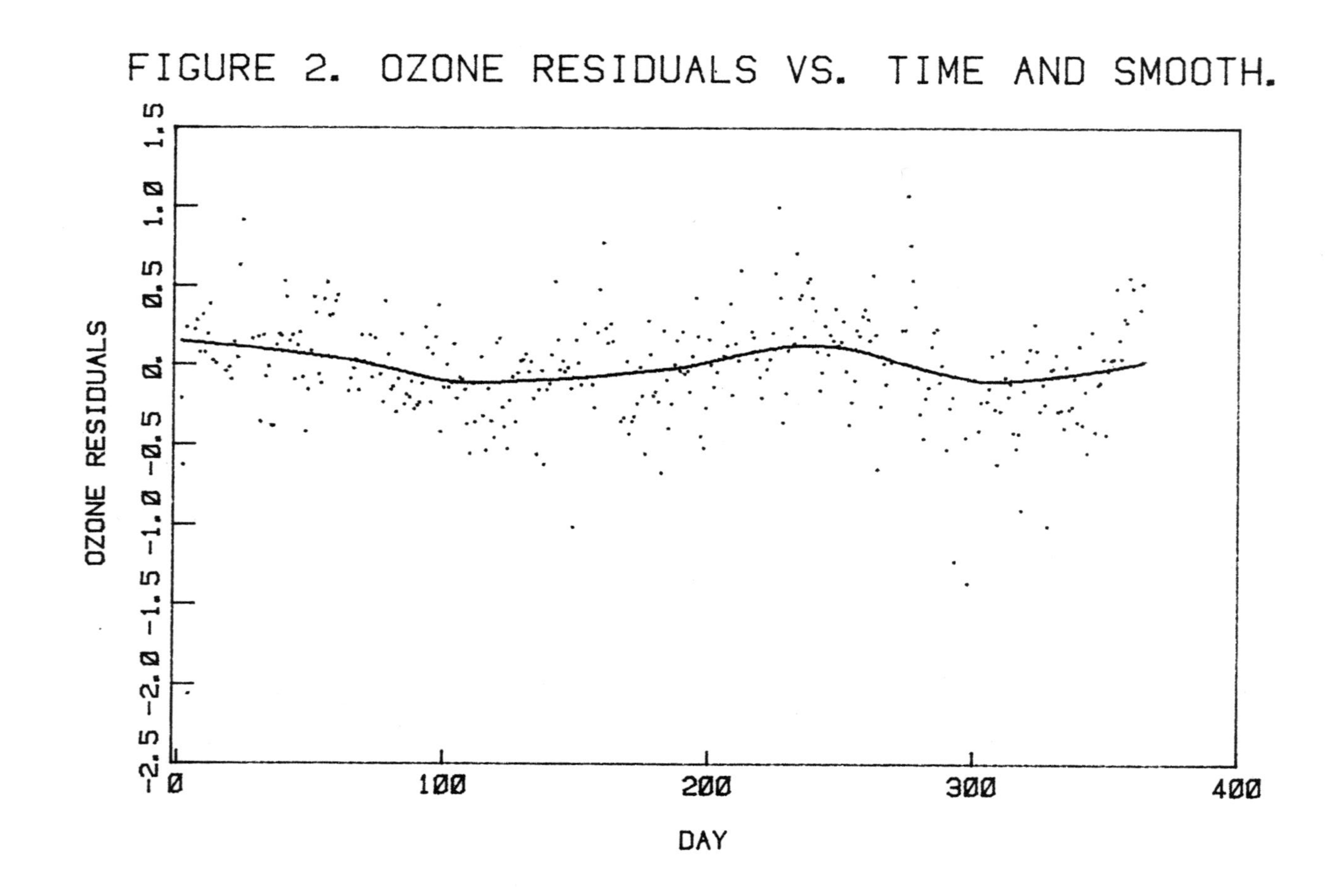

FIGURE 2. OZONE RESIDUALS VS. TIME AND SMOOTH.

(7) If a robust estimation procedure is used, such as M-estimation with the bisquare (Mosteller and Tukey, 1977) or the Huber (Huber, 1973) weight function, a symmetry plot of the residuals should be made to check the symmetry of the error terms.

(8) How well the equation fits the data can be portrayed by a plot of the empirical cumulative distribution function of the residuals, a plot of the response against the fitted values, and by the plot in (4).

The following references are given to guide the reader toward discussions relating to these eight points.

3.1 Point (1)

Daniel and Wood (1971) and Wood (1973) make this point both forcefully and convincingly. One solution is to attempt using multi-dimensional graphics. The method being developed in (Friedman, Jacobson, and Stuetzle, 1980) and (Friedman and Stuetzle, to appear) is a step in this direction.

3.2 Point (2)

Multi-dimensional plotting techniques have been surveyed in many places including (Gnanadesikan, 1977) and (Chambers and Kleiner, to appear).

3.3 Point (3)

I do not know much about the history of this plot. It appears in (Box et al., 1957) but little is made of it. Gnanadesikan advocated its regular use as discussed by Larsen and McCleary (1972a); unfortunately in the published version (Larsen and McCleary, 1972b) the discussion was deleted by somewhat overzealous editing by the referees. More recently, Mosteller and Tukey (1977) discuss this plot. A related procedure, sometimes called partial residual plots, was devised by Ezekiel and further discussed by Larsen and McCleary (1972b) and Wood (1973). To this author, the adjusted variable plot seems more helpful than the partial residual plot.

3.4 Point (4)

This classical plot (apart from the smoothing) is discussed by Anscombe and Tukey (1963), Daniel and Wood (1971), and Draper and Smith (1966).

3.5 Point (5)

Several examples illustrating the problem and the solution are given in (Cleveland, 1979).

3.6 Point (6)

Not only do Daniel and Wood (1971) advocate this procedure but they also provide many examples of normal probability plots of data that are normally distributed, so the reader can accustom his eyes to the look of the plot in the null hypothesis case.

3.7 Point (7)

Symmetry plots are discussed in Wilk and Gnanadesikan (1968).

3.8 Point (8)

There are many good discussions of these and other points by Mosteller and Tukey (1977). The empirical cumulative distribution function plot is discussed by Wilk and Gnanadesikan (1968).

4. Computing

A routine called LOWESS has been written for smoothing scatterplots by robust locally weighted regression. The graphical methods for regression described here can be easily carried out using the graphics package GR-Z. Information on obtaining LOWESS or GR-Z may be gotten from Computing Information Library, Bell Laboratories, 600 Mountain Ave., Murray Hill, N.J. 07974.

References

Anscombe, F. J. and Tukey, John W. (1963). The examination and analysis of residuals. *Technometrics* **5**, 141-160.

Box, G. E. P., Cousins, W. R., Davies, O. L., Hinsworth, F. R., Henney, H., Milbourn, M., Spendly, W., and Stevens, W. L. (1957) *Statistical Methods in Research and Production,* Third Edition, Oliver and Boyd, London.

Chambers, John M., Cleveland, William S., Kleiner, Beat, and Tukey, Paul A. (to appear). *Graphical Methods for Data Analysis.*

Chambers, John M. and Kleiner, Beat (to appear). Graphical Techniques for Multivariate Data and for Clustering. *Handbook of Statistics* II, edited by Krishnaiah. North-Holland, New York.

Cleveland, William S. (1979). Robust Locally Weighted Regression and Smoothing Scatterplots. *Journal of the American Statistical Association* **74**, 829-836.

Daniel, Cuthbert and Wood, Fred S. (1971). *Fitting Equations to Data.* Wiley, New York.

Draper, N. R. and Smith, H. (1966). *Applied Regression Analysis.* Wiley, New York.

Friedman, Jerome H., Jacobson, Mark, and Stuetzle, Werner (1980). *Technical Report No. 146,* Dept. of Statistics, Stanford Univeristy.

Friedman, Jerome H. and Stuetzle, Werner (to appear). Projection Pursuit Methods for Statistical Data Analysis. *ARO Workshop on Modern Data Analysis.*

Gnanadesikan, R. (1977). *Methods for Statistical Data Analysis of Multivariate Observations.* Wiley, New York.

Horowitz, Joel (to appear). Extreme Values from a Nonstationary Stochastic Process: An Application to Air Pollution Analysis. *Technometrics.*

Huber, Peter J. (1973). Robust Regression: Asymptotics, Conjectures, and Monte Carlo. *Annals of Statistics* **1**, 799-821.

Larsen, Wayne A. and McCleary (now Devlin), Susan J. (1972a). The Use of Partial Residual Plots in Regression Analysis. Uneditted version of next reference. Available from the second author at Bell Laboratories, 600 Mountain Ave., Murray Hill, N.J. 07974.

Larsen, Wayne A. and McCleary (now Devlin), Susan J. (1972b). The use of partial residual plots in regression analysis. *Technometrics* **14**, 781-790.

Mosteller, Frederick and Tukey, John W. (1977). *Data Analysis and Regression.* Addison-Wesley, Reading, Massachusetts.

Stone, Charles J. (1977). Consistent Nonparametric Regression, *Annals of Statistics* **5**, 595-620.

Wilk, M. B. and Gnanadesikan, R. (1968). Probability Plotting Methods for the Analysis of Data. *Biometrika* **55**, 1-17.

Wood, Fred S. (1973). The use of individual effects and residuals in fitting equations to data. *Technometrics* **15**, 677-695.

SOME COMPARISONS OF BIPLOT DISPLAY AND PENCIL-AND-PAPER E.D.A. METHODS[1]

Christopher Cox
K. Ruben Gabriel

Department of Statistics and
Division of Biostatistics
University of Rochester
Rochester, New York

Presented at ARO Workshop on Modern-Data Analysis at
Raleigh, North Carolina
June 2-4, 1980

This paper presents some comparisons of EDA and biplot display. By pencil-and-paper EDA we mean the methods advocated by John Tukey in his 1977 volume (Tukey, 1977). We use examples from that book to illustrate the differences and the similarities of the two methods. We assume that the analyses in the book are familiar and show how our biplot analyses differ from them.

The paper begins with an introduction to the biplot, accompanied by one example, in which the biplot is used for data summarization and description. Then we look at four diagnostic examples from the book and show what biplot display would have done. We end by drawing some conclusions. References for further reading on the biplot and its diagnostic and

[1]Research supported in part by ONR contract N 00014-80-C-0387 on Biplot Multivariate Graphics.

ISBN 0-12-438180-4

other uses are included at the end of the paper. Computer programs are available from the authors at the Division of Biostatistics of the University of Rochester.

We start by explaining what the biplot is. It is a graphical display of a matrix $Y_{(nxm)}$ of n rows and m columns by means of row markers $\underline{a}_1, \underline{a}_2, \ldots \underline{a}_n$ and column markers $\underline{b}_1, \underline{b}_2, \ldots \underline{b}_m$. The biplot carries one marker for each row, and one marker for each column. The principle of biplot display of a matrix Y is that element $y_{i,j}$ in the i-th row and j-th column is represented by the inner product of the i-th row marker and the j-th column marker, i.e., $\underline{a}_i'\underline{b}_j$ represents $y_{i,j}$. A 100 by 20 matrix, for example, would be represented by 100 row markers and 20 column markers in such a way that all 2,000 elements are represented by inner products of row markers and column markers. To set this in matrix terms we may array the row markers $\underline{a}_i$ as rows of matrix A and the column markers $\underline{b}_j$ as columns of a matrix B'. Clearly, then the matrix product AB' represents the matrix Y itself.

On a point of terminology, the prefix "bi-" of biplot serves to indicate that this is a joint display of rows and columns. It does not indicate the two-dimensionality of the biplot. Any plot is two dimensional. On the other hand, if we use a three-dimensional display analogous to the biplot, we call it a bi-model because it too is a joint display of both rows and columns: the ending "model" indicates that there are three dimensions.

Figure 1 shows a very simple example of a biplot. Y is a 4 x 3 matrix of rank 2; the row markers are the rows of matrix A, and the column markers are the columns of matrix B'. Each row of A and each column of B' is displayed on this biplot--

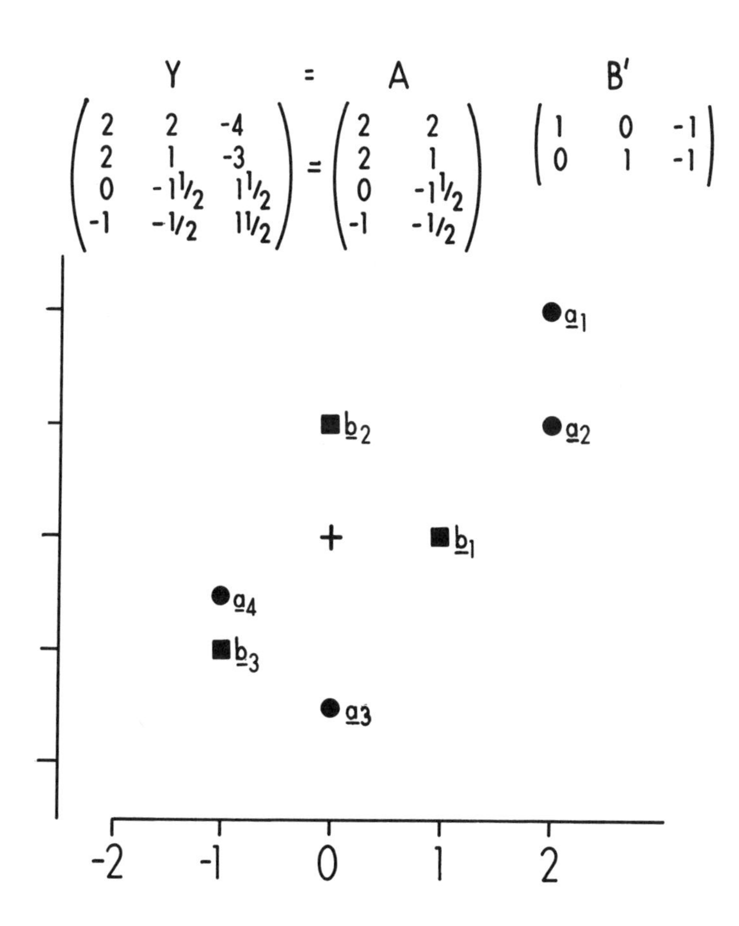

Fig. 1. A matrix Y, its factorization AB' and the biplot.

seven markers in all. For convenience the row markers $\underline{a}_i$ are indicated as circles whereas the column markers $\underline{b}_j$ are indicated as squares.

Figure 2 illustrates how particular elements of the matrix are represented on the biplot. Thus, element $y_{2,3}$ is represented by the inner product of the second row marker $\underline{a}_2$ and the third column marker $\underline{b}_3$. To see this geometrically, we choose one of these markers, e.g., $\underline{b}_3$, take the straight line

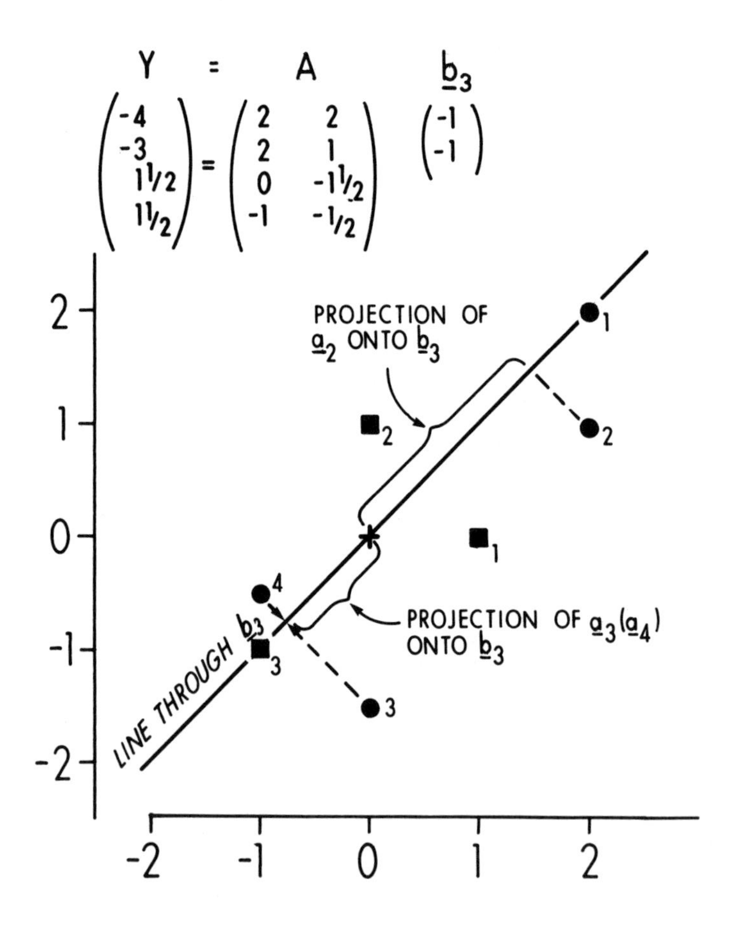

Fig. 2. Biplot representation of the third column of Y.

from the origin through that marker, and project the other marker $\underline{a}_2$ orthogonally onto it. The distance from the origin to the foot of the perpendicular of $\underline{a}_2$ onto the line through $\underline{b}_3$ is then multiplied by the length of the vector $\underline{b}_3$ to obtain the inner product. For another example, take $y_{3,3}$ for which we project $\underline{a}_3$ onto $\underline{b}_3$. In this case, the projection is half as long as the one before and in the opposite direction, i.e., in the direction of $\underline{b}_3$ itself. So the product is positive and half the size of the previous one.

A few more remarks about biplots need to be made. First of all note that the biplot is planar--the row markers $\underline{a}_i$, as well as the column markers $\underline{b}_j$, are plotted in the plane. This cannot be done exactly for any matrix of rank greater than 2. Hence, the first step for biplotting such a matrix Y is to approximate it by a matrix $Y_{[2]}$ of rank 2. This is called lower rank approximation. The second step is to factorize the $Y_{[2]}$ approximation into a product AB' of an A matrix of two columns and a B' matrix of two rows. Then the rows of A can be plotted as row markers $\underline{a}_i$ and the columns of B' as column markers $\underline{b}_j$. Their joint display is a biplot. Biplotting is thus seen to require three steps: rank 2 approximation, factorization, and display.

Lower rank approximation can be carried out by means of the theorem due to Householder and Young (1938), which provides the least squares solution to this problem. When weights are introduced, and each of the squared differences $(y_{ij} - \underline{a}_i'\underline{b})^2$ is to be weighted by some given w_{ij}, the mathematics of that theorem break down. However, a weighted least squares algorithm and suitable initialization methods are available (Gabriel and Zamir, 1979). An earlier solution for particular kinds of weights common in statistics was provided by Haber (1975). Another method of reduced rank approximation uses adaptive fits (McNeil and Tukey, 1975). C. L. Odoroff of Rochester is currently working on using the weighted least squares solution for adaptive fitting, i.e., taking the residuals from the last fit and using them to adjust the weights for the next fit.

We now turn to uses of the biplot. These are mostly of two kinds--inspection of data and diagnostics. We present one brief example of biplot inspection before we go on to our main subject which is biplot diagnostics.

We consider data from single-dose, postoperative, oral analgesic trials. Patients who had previously consented to participate, and who requested medication for moderate to severe pain during the first three days after surgery, were given a single dose of one of the study drugs on a randomized, double-blind basis. The resulting data consist of ordinal pain scores, on a five point scale, with zero being no pain and four being very severe pain. (A number of standard pain scales were used, of which we have chosen this one as an illustration.) Data were recorded at baseline (medication time), one-half hour later, and at hourly intervals until five hours after medication--Figure 3. There were a total of 180 patients in the trials, with eight treatments, including a placebo.

In Figure 4 we show a portion of a biplot of the data matrix. We have included only two of the treatment groups: those receiving a placebo and those given a highly effective

Surgical Patients	Baseline (Medication Time)	Time after medication					
		1/2 hr.	1 hr.	2 hrs.	3 hrs.	4 hrs.	5 hrs.
1	. . . .	. .	. .	. .	. .	. .	. .
2	. . . .	. .	. .	. .	. .	. .	. .
				DATA			
180	. . . .	. .	. .	. .	. .	. .	. .

Note: Ordinal pain scores on a 5-point scale from 0 (no pain) to 4 (worst pain I have ever experienced).

Fig. 3. Data from single dose postoperative oral analgesic trials.

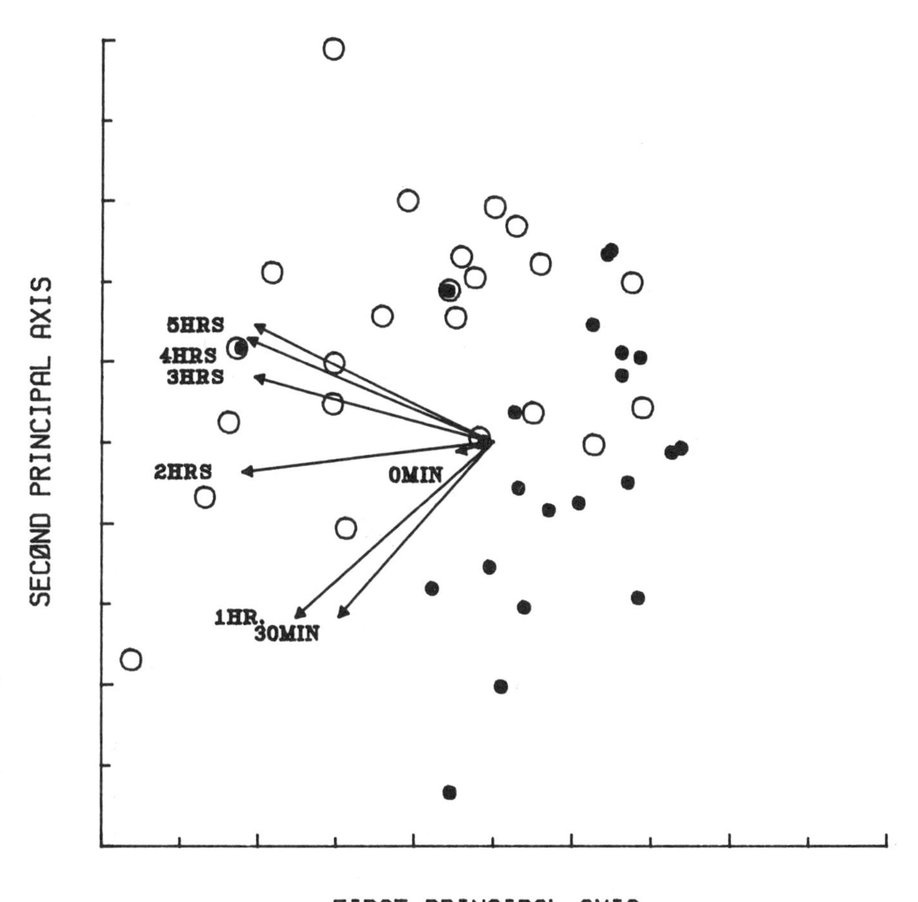

Fig. 4. Biplot of ordinal pain scores from two treatment groups in analgesic trials.

combination analgesic (pentazocine and aspirin). This particular biplot has been scaled so that the lengths of the arrows represent standard deviations and the angles between the arrows represent correlations among the corresponding columns--times of recording. (This is referred to as a GH' biplot--Gabriel, 1971.)

Figure 5 replaces the row markers of Figure 4 by one standard deviation concentration ellipses. Each ellipse summarizes the row markers of the approximately 25 patients in the corresponding treatment group. The center of each ellipse is also plotted.

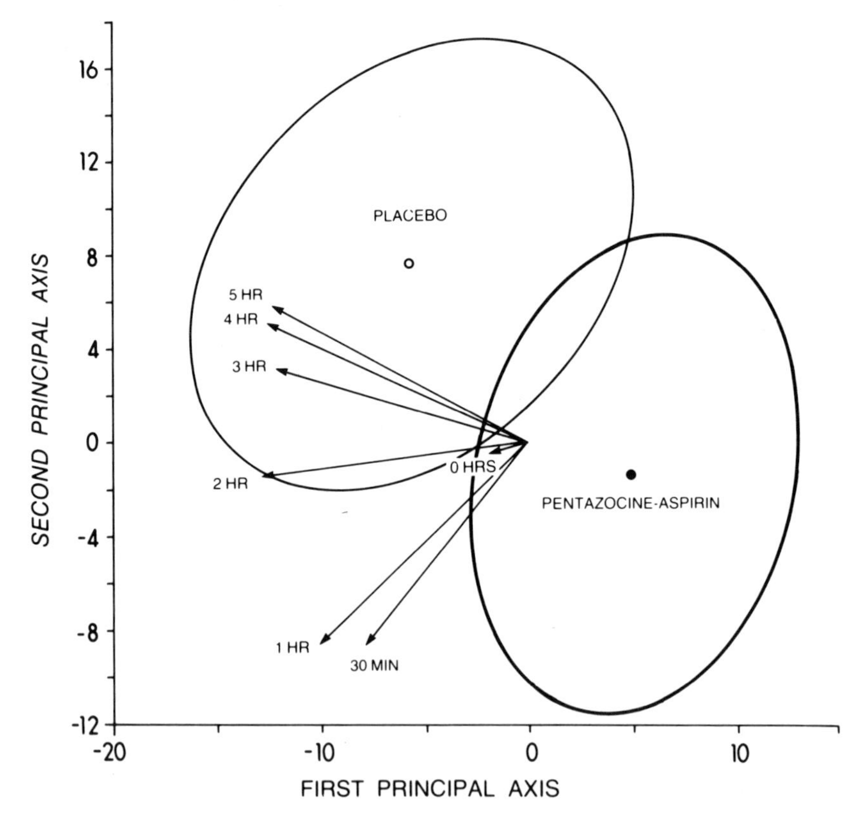

Fig. 5. Analgesic biplot with concentration ellipses summarizing treatment groups.

The small angle between the arrows for the one-half hour and one hour pain scores shows that scores at those times are fairly well correlated with one another. Similarly, one can see that the three, four, and five hour pain scores are correlated with each other, but roughly uncorrelated--arrows at about 90°--with the scores at earlier times. The shortness of the baseline arrow reflects the fact that only patients with baseline pain scores of 2 or 3 were included in these trials. (Further examination of the data suggests that the baseline pain scores are not well represented by the biplot.) In order to see the effects of the two treatments, we examine the average pain scores of the two treatment groups at different time points by projecting the centers of the ellipses onto the arrows. We see that both groups were similar (and below the overall average) at the one-half and 1 hour time points; this is an indication of a placebo effect of surprising duration. With respect to the later time points, however, the placebo group had much higher than average pain scores, while the pentazocine-aspirin group had below average pain scores. (The other six treatment groups were intermediate, increasing in efficacy from placebo.) Clearly, the later time points are more sensitive to the effect of analgesics. This analysis contrasts with the traditional method of analyzing such data in which the post-medication pain scores are cummulated to obtain a measure of "total pain" and the time differentials are ignored (Cox et al, 1980). Inspection of this biplot has suggested new derived pain measures, which appear to be more sensitive to treatment differences.

We now turn to the second use of the biplot which is to facilitate the search for patterns and the inference of models to fit the data. To illustrate, some of the patterns that we look for are shown in Figure 6. Display 6a shows row markers and column markers which are both collinear and have a right angle between their two lines. Such a biplot pattern indicates that the data are well fitted by an additive model, i.e., $y_{i,j} = \alpha_i + \beta_j$ for some row effects α_i and some column effects β_j. This is something that the eye picks up readily: row markers on a line, column markers on a line, and a 90° angle between them.

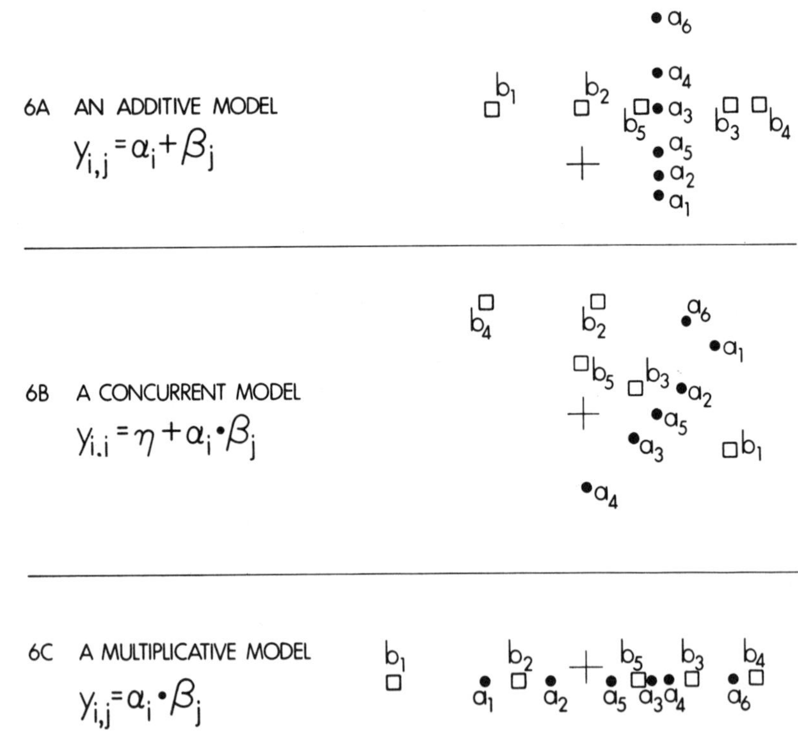

Fig. 6. Biplot patterns and the models which they diagnose.

Another biplot pattern, shown in Display 6b, has the column markers aligned, and the row markers aligned, but the angle between the lines is not 90°. A concurrent model can then be diagnosed. This is also known as a degree of freedom for non-additivity model and it can be parametized most simply as $y_{i,j} = \eta + \alpha_i \beta_j$.

Finally, Figure 6c has all markers on one line. This obviously diagnoses a model of rank one, i.e., $y_{i,j} = \alpha_i \beta_j$.

It will have been noticed that in Figure 6b there is one column marker--$\underline{b}_3$--that is not aligned with the other $\underline{b}$'s and one row marker--$\underline{a}_4$--that is not aligned with the other $\underline{a}$'s. That indicates that the third column and fourth row are not fitted by the concurrent model, though the other rows and columns are. This illustrates a very useful property of the biplot; if a pattern fits only some of the column markers and some of the row markers, the implied model may be diagnosed exclusively for those columns and rows. Indeed, a biplot is not only a display of the whole matrix, but can also be regarded as a simultaneous display of all possible submatrices. The eye immediately picks up subsets and subtables and allows their separate diagnosis.

We should add that outlying rows or columns might at times distort the rank two approximation and spoil the chances of diagnosing a model. There might also be situations where the subtable models cannot be seen on the biplot because the biplot mainly displays subtable differences. In such cases it might be helpful to employ a 3D bimodel and see whether any simple patterns are evident on some projection of such a bimodel.

Row Markers $\underline{a}_i$	Column Markers $\underline{b}_j$	The Model for $y_{i,j}$ is:
Collinear	-	$\beta_j + \alpha_i\delta_j$ columns regression
-	Collinear	$\alpha_i + \gamma_i\beta_j$ rows regression
Collinear	Collinear	$\mu + \gamma_i\beta_j$ one degree of freedom for non-additivity
Collinear lines 90° to each other	Collinear	$\alpha_i + \beta_j$ additive

Fig. 7. Some biplot diagnostic rules (Bradu and Gabriel, 1978).

The examples of Figure 6 illustrate some simple diagnostic rules which are listed more formally in Figure 7. There are four collinearity patterns for row and/or column markers and Figure 7 shows the model that may be diagnosed from each one. Thus, collinear row markers indicate that each column can be modelled by a linear regression on some "row effects" α_i. An analogous diagnosis follows from column marker collinearity.

TABLE I. Monthly Mean Temperatures (°s F)[a]

	Caribou	Washington, D.C.	Laredo
Jan.	8.7	36.2	57.6
Feb.	9.8	37.1	61.9
Mar.	21.7	45.3	68.4
Apr.	34.7	54.4	75.9
May	48.5	64.7	81.2
Jun.	58.4	73.4	85.8
Jul.	64.0	77.3	87.7

[a]From J.W. Tukey, EDA, Chapter 10; and Climatography of the United States, (1959-60), Numbers 60-17, 60-18, and 60-41, Washington, D.C.: U.S. Weather Bureau.

Joint collinearity diagnoses concurrence or additivity, depending on the angle between the lines, as already illustrated in Figure 6.

We now turn to the first example from John Tukey's book. Table I shows monthly mean temperatures at three locations, one up North, one mid-way and one further South. The biplot of the data is shown in Figure 8. The biplot column markers for the three cities are clearly collinear and the row markers for months are also pretty close to collinear. The angle between the lines is not 90°, and this suggests a concurrent

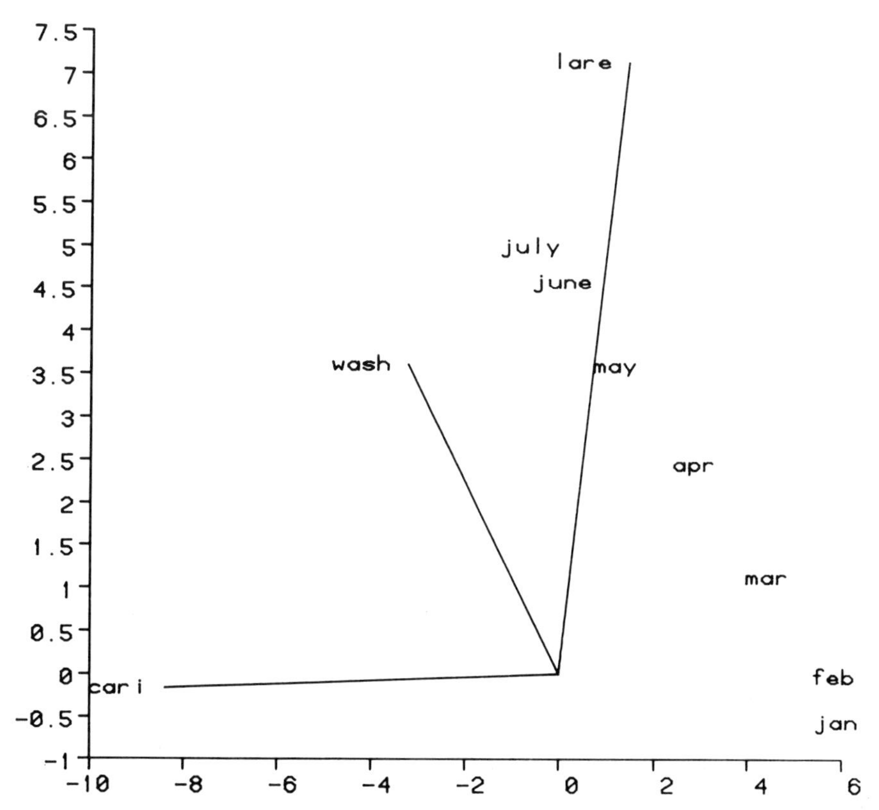

Fig. 8. Biplot of mean monthly temperature data from three cities (Caribou; Washington, D.C.; Laredo).

model. Indeed, that is exactly what Tukey concluded in his book, where he calls it a "plus-one-fit".

It is evident that the biplot has revealed this model very simply and strikingly. Actually, a few more things may be said about this example. The months really are not quite collinear--they seem to curve around in a sequence from January to July. This leads one to wonder about addition of the remaining months. Tsianco (1980) has done similar biplots on data of 50 weather stations for twenty-four successive months. When one looks at part of his biplots, they look much like Figure 8, but when one looks at a bimodel of the entire years' data, the month markers are found to be on an ellipse in 3D. It can be shown that an ellipse on the biplot diagnoses a harmonic model for the data. That is a more reasonable model for temperature than a "concurrent", or plus-one-fit, model.

This example is considered again in Chapter 9 of Mosteller and Tukey (1977), where a one parameter family of "matched" exponential transformations is used essentially to obtain additivity (e.g., d = 70 in Exhibit 21). The biplot of the transformed data, however, still suggests a concurrent model, even though the median polish residuals do not suggest this. Thus, the biplot can serve as a useful check whether a transformation has achieved its purpose.

As a second example of the use of these diagnostic rules, we consider data on the world's supply of telephones -- Table II A--analyzed in Tukey's (1977), Chapter 12. The world was divided into seven "continents", and yearly counts were given from 1951 to 1961, with the years 1952 through 1955 omitted. Yearly increases are seen to be more proportional

TABLE II A. World's Telephones (Raw Counts).[a]

	1951	1956	1957	1958	1959	1960	1961
N.Amer.	45939	60423	64721	68484	71799	76036	79831
Eur.	21574	29990	32510	35218	37598	40341	43173
Asia	2876	4708	5230	6062	6856	8220	9053
S.Amer.	1815	2568	2695	2845	3000	3145	3338
Oceania	1646	2366	2526	2691	2868	3054	3224
Africa	895	1411	1546	1663	1769	1905	2005
MidAmer.	555	733	773	836	911	1008	1076

[a]From J. W. Tukey, *EDA*, Chapter 12; and *The World's Telephones*, (1961), American Telephone and Telegraph Co.

TABLE II B. Log_e Counts of World's Telephones

	1951	1956	1957	1958	1959	1960	1961
N.Amer.	10.735	11.009	11.078	11.134	11.182	11.239	11.288
Eur.	9.979	10.309	10.389	10.469	10.535	10.605	10.673
Asia	7.964	8.457	8.562	8.710	8.833	9.014	9.111
S.Amer.	7.504	7.851	7.899	7.953	8.006	8.054	8.113
Oceania	7.406	7.769	7.834	7.898	7.961	8.024	8.078
Africa	6.797	7.252	7.343	7.416	7.478	7.552	7.603
MidAmer.	6.319	6.597	6.650	6.729	6.815	6.916	6.981

than additive, and we follow Tukey's suggestions and consider the logarithms--Table II B.

We first examine a biplot of the mean-centered log counts, shown in Figure 9. In addition to plotting the row and column markers, we have also included their arithmetic averages gmn and hmn, for row and column markers, respectively. From the evident collinearity of the column markers we diagnose a rows regression model--second row of Figure 7. The linearity of regression on time is checked by comparing the distances between column markers with the corresponding time intervals. It is thus evident from the figure that the regression is linear in time.

We next show how to use the biplot to obtain approximate parameter estimates, and thus more specific diagnoses. (For details see Bradu and Gabriel, 1978.) We first draw the line through the column markers and project the row markers orthogonally onto it. The distances from these projections to the projection of the mean of the row markers (gmn) are proportional to the estimates of the regression coefficients γ_i. (The projection of the mean gives the positive direction.) On the basis of these projections, we decide to fit a single slope for North, South, and Mid-America, and Oceania. We also require the same slope for Europe as Africa. A higher slope is clearly needed for Asia.

Further diagnosis can be obtained by projecting the row markers onto the line through the origin and the mean of the column markers (hmn). Distances from the projection of the mean (gmn) approximate the row effects (α_i) proportionally. These are the intercepts of the regressions. Indeed, the ordering of these projections is quite similar to that of the log counts in 1951. Thus, for example, North America had more telephones than Europe in 1951, but subsequently Europeans acquired them more rapidly.

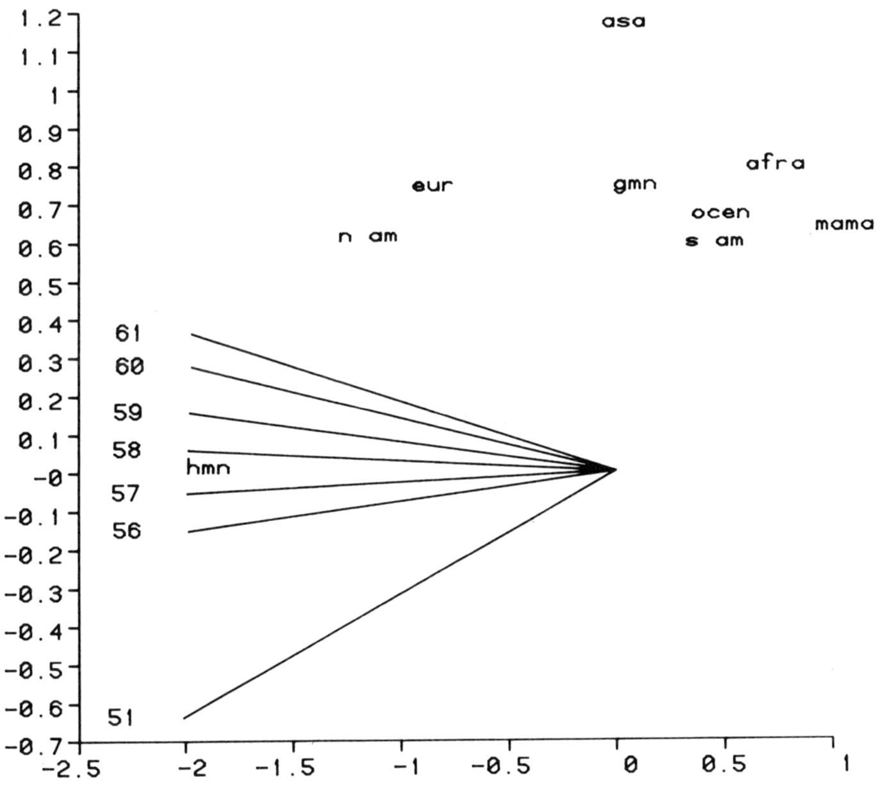

Fig. 9. Biplot of log counts of telephones, by continent and year.

TABLE III A. Least Squares Fit to Logs of Telephone Data.

	Regression Coefficients	
	Intercept	Slope
N.Amer.	10.63	0.063
Eur.	9.86	0.076
Asia	7.80	0.116
S.Amer.	7.45	0.063
Oceania	7.39	0.063
Africa	6.79	0.076
MidAmer.	6.25	0.063
		GOF = 0.9997

TABLE III B. Least Squares Fit to Residuals for Years 1956-61.

	Intercept	Slope
N.Amer.	0.0567	-0.0075
Eur.	0.0207	-0.0033
Asia	-0.1751	0.0197
S.Amer.	0.0887	-0.0103
Oceania	0.0089	-0.0002
Africa	0.0607	-0.0059
MidAmer.	-0.1492	0.0174
		GOF = 0.9999

Table III A shows the results of our first least squares fit. The goodness-of-fit, expressed as the sum of squared residuals divided by the sum of squared deviations from the overall mean, is very good (99.97%), and we are tempted to stop here, or perhaps fit a model with fewer intercept parameters. (This fit is similar to the fit displayed in Exhibit 14 A--Chapter 12--of Tukey's book, if one improves it slightly by adjusting the row slope for Mid-America to be equal to 30.)

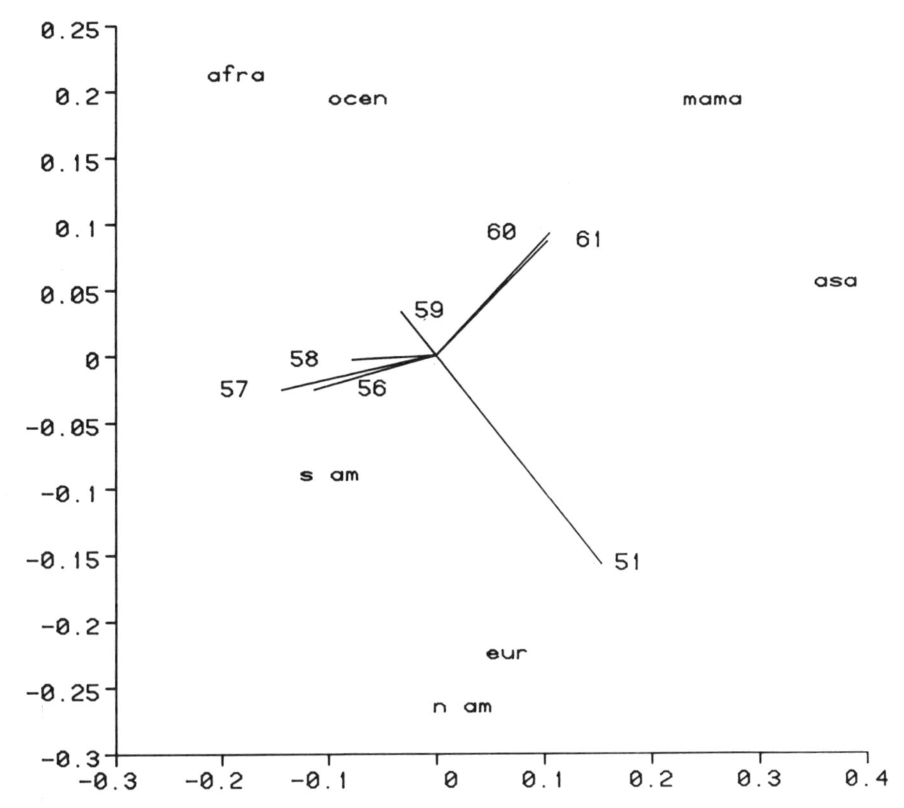

Fig. 10. Biplot of residuals from first fit to log telephone counts.

In Figure 10 we take a second look and biplot the residuals from our first fit. This biplot still shows a trace of collinearity in the column markers from 1956 or 1957 to 1961, and that would diagnose a rows regression model for a subtable. Table III B shows the estimates of the linear regression parameters (one line for each continent) for the 1956-61 subtable. We see that, for Asia and MidAmerica, our first fit overestimated the rates of increase in numbers of telephones from 1951 to 1956, and underestimated them afterwards. The opposite is true for the remaining continents. In general, the year 1951 probably had too great an influence on our first fit. The extra fit makes a small improvement in goodness-of-fit. However, we consider this less important than the extra information we have obtained about time changes in telephone acquisition.

A biplot of the residuals from this additional fit, shown in Figure 11, reveals much less structure than the previous one, although some regularity remains. In this respect one is reminded of the famous "vapor pressure of water" example (Tukey, 1977, Chapter 6) in which definite structure remains in the residuals even after a clearly diagnosed fit.

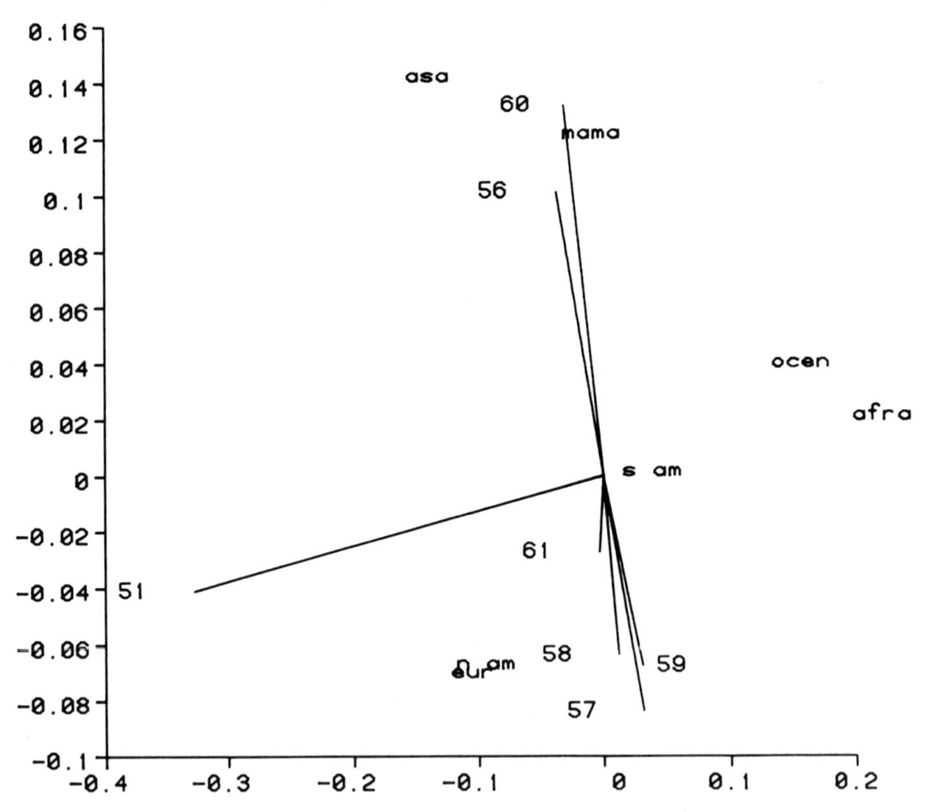

Fig. 11. Biplot of residuals from extra fit to years 1956 to 1961.

Our next example concerns the data in Table IV, which are from an experiment for measuring the sensitivity of several

TABLE IV A. Finger Limens.[a]

	Initial Weights	Persons	Rates r_a	r_b	r_c	r_d
Raw	W_1	K	39	85	101	151
		L	16	32	43	63
		M	18	31	42	58
	W_4	K	31	55	84	124
		L	12	22	33	51
		M	13	26	38	55
	W_7	K	26	56	70	98
		L	12	20	30	37
		M	14	26	40	46

[a]From J. W. Tukey, EDA, Chapter 13; and Table 89 of Johnston, P.O., (1949) Statistical Methods in Research, New York: Prentice-Hall.

TABLE IV B. Logs of Finger Limens.

Initial Weights	Persons	Rates r_a	r_b	r_c	r_d
W_1	K	3.664	4.443	4.615	5.017
	L	2.773	3.466	3.761	4.143
	M	2.890	3.434	3.738	4.060
W_4	K	3.434	4.007	4.431	4.820
	L	2.485	3.091	3.497	3.932
	M	2.565	3.258	3.638	4.007
W_7	K	3.258	4.025	4.248	4.585
	L	2.485	2.996	3.401	3.611
	M	2.639	3.258	3.689	3.829

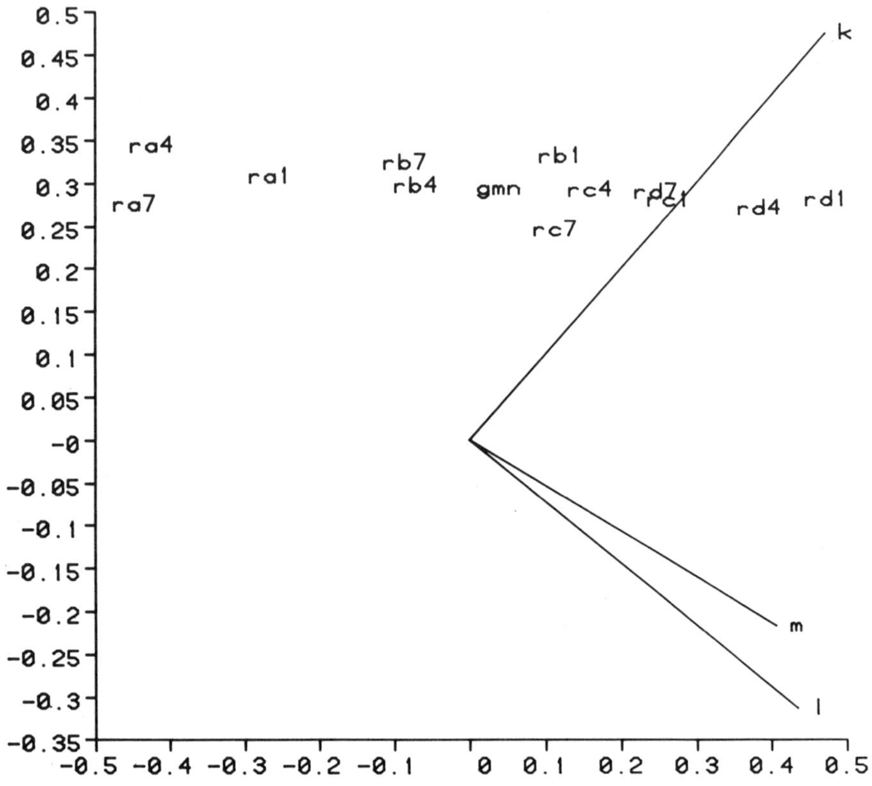

Fig. 12. Biplot of logs of finger limens: rates and weights vs. individuals.

individuals to changes in pull, and are analyzed in Chapter 13 of Tukey's (1977) book. The table shows data for individuals K, L and M, for different initial steady pulls W_1, W_4, W_7, which are referred to as "weights", and different rates of increase of the pull, r_a,r_b, r_c and r_d. We follow Tukey's suggestion and analyze the logs shown in Table IV B.

The first problem with displaying these data is that they are in a three-way layout. As the biplot is a matrix, i.e., two-way, display, it can be applied to these data only if two of the three classifications are crossed either in the rows or in the columns of a matrix--as in Table IV in which weights and individuals are crossed in the rows. There are two other ways of crossing in the rows (there are also three transpositions with crossing in the columns--but their biplots do not differ from the previous three). It will be instructive to look at all three biplots. (Kester (1979) has considered biplot display of such three- and higher-way layouts.)

Figure 12 shows the biplot with individuals in the columns and the rates and weights crossed in the rows. At first it is a little difficult to see any pattern because there are too many row markers. But if one pencils in lines to join the three weights for each rate, a clear pattern emerges. The average of the r_a markers is farthest to the left, then the average of the r_b markers, then that for r_c and finally, the average for r_d--and those four averages are more or less collinear. Furthermore, this line of averages is approximately at right angles to a line through the three markers for individuals. Thus the rate classification appears orthogonal to that by individuals. According to the rules in Figure 7, this suggests additivity between rates and individuals.

From this figure, one can also infer the relative sizes of the differences between rates and weights. In this biplot the positive direction is to the right, as the arrows point there (recall the inner-product construction). The order of the rates in that direction is $r_a < r_b < r_c < r_d$. The order of weights is $W_7 < W_4 < W_1$, but the average differences between the weights are much smaller than those between the rates. We will not discuss such comparisons in detail but rather focus on model diagnoses.

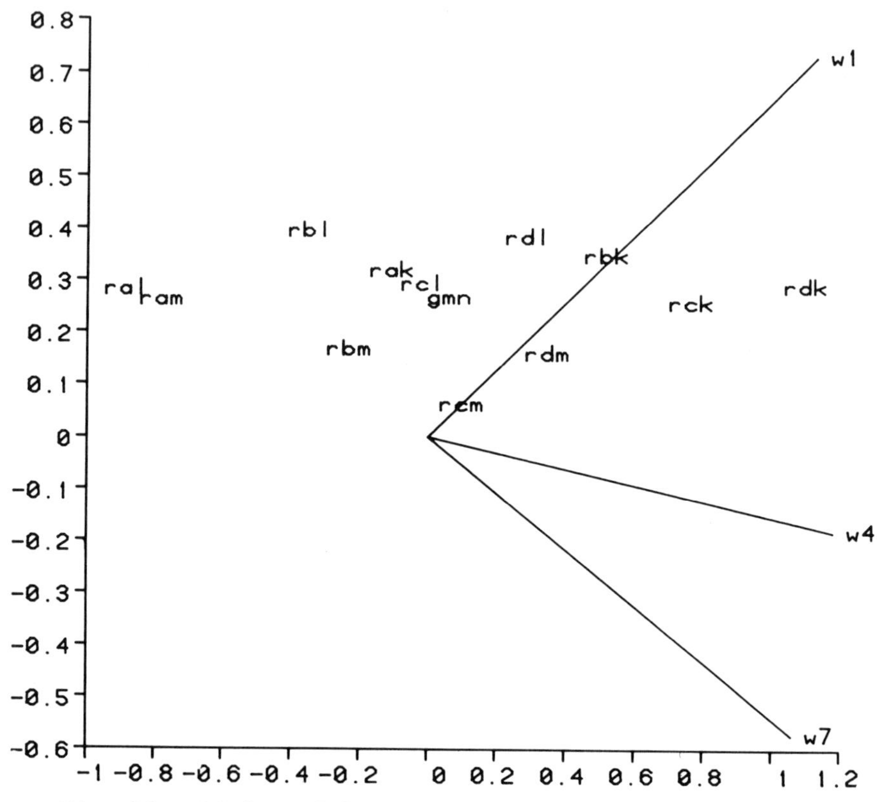

Fig. 13. Biplot of logs of finger limens: rates and individuals vs. weights.

Figure 13 has individuals crossed with rates in the rows. The column markers for weights are pretty much collinear. The row markers seem a bit messy but if one looks at them carefully or draws a few lines, one sees that for each individual the rates are close to a straight line orthogonal to the direction of the weights' line. The diagnosis would therefore be that rates are additive with weights.

From Figure 12, we have rates additive with individuals; from Figure 13, we have rates additive with weights. Together, these diagnoses indicate a model $y_{krw} = \alpha_r + \Theta_{kw}$, in which the variable y is indexed by individual k, rate r, and weight w. The model has a rate effect α_r which is additive to a joint weight-individual effect θ_{kw}, which allows weight-individual interaction. Indeed, in Figure 13, K, L and M are not collinear, so there is no reason to expect individuals to be additive with weights. It is easy to see that the interaction is due mostly to individual M: the weight markers line is seen to be pretty much orthogonal to K averages, but the average of the M markers is not on that line. The interaction is thus seen to consist mainly of individual M's having a relatively large value for weight 7 and a relatively small value for weight 1. These finding are very similar to those in the analysis in Tukey's (1977) book.

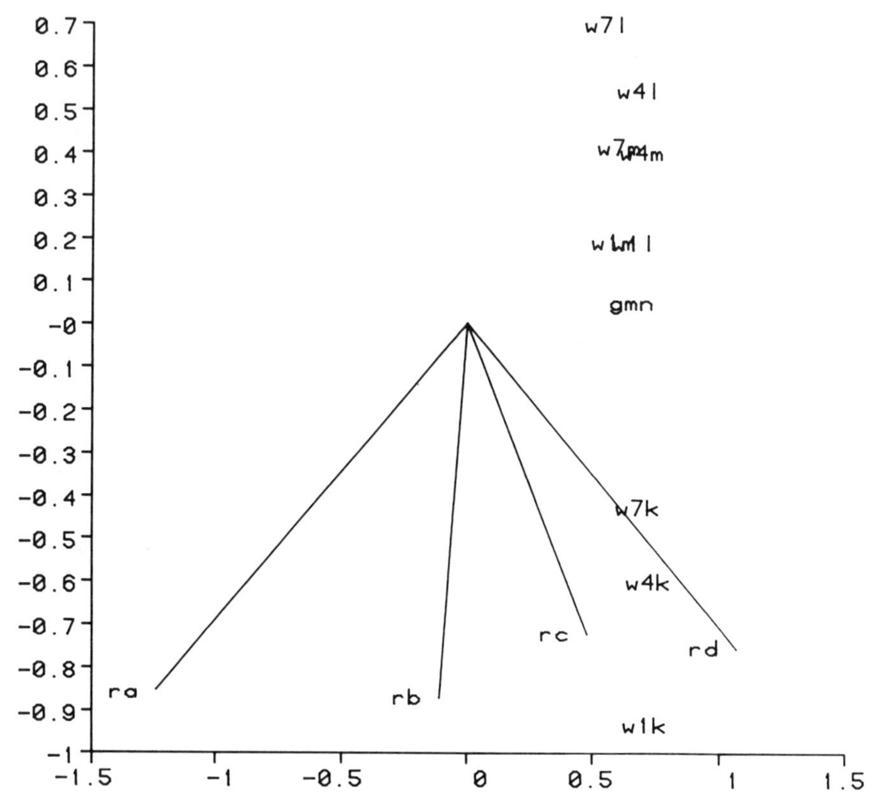

Fig. 14. Biplot of logs of finger limens: weights and individuals vs. rates.

The third biplot is shown in Figure 14. This shows a rather more striking feature than either of the previous two. The rates are represented by fairly collinear column markers; individuals and weights are represented together by collinear row markers at a nearly right angle to the line which roughly fits the rate markers. The most striking feature of this figure is that weights and individuals markers are jointly collinear. By the first rule of Figure 7, the model is diagnosed as $y_{krw} = \alpha_r + \beta_r \Theta_{kw}$--a regression of the rates onto

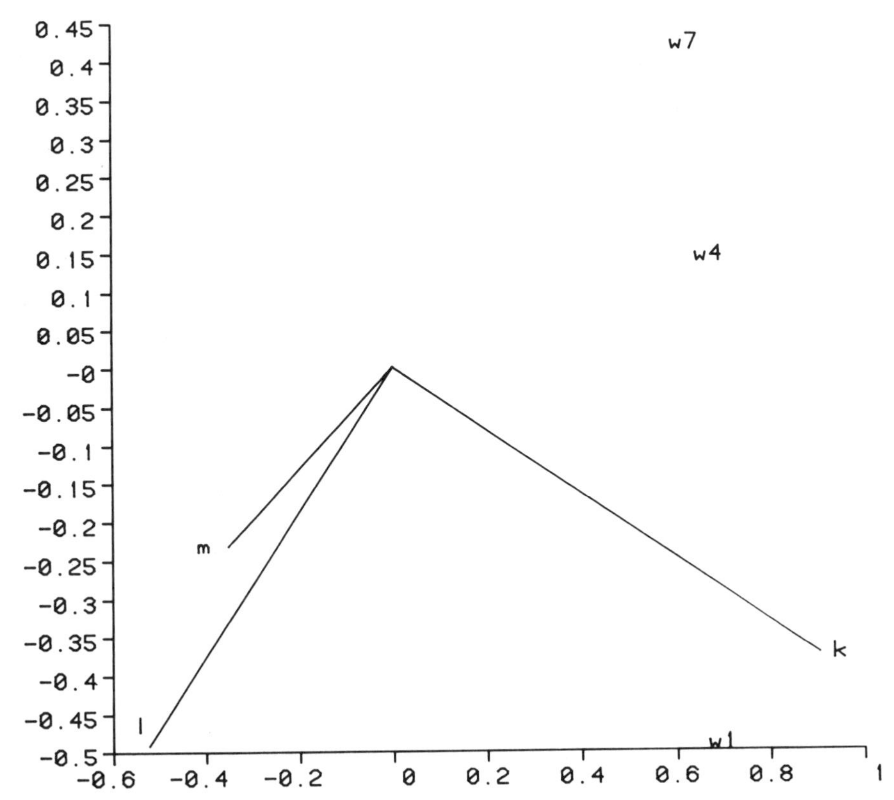

Fig. 15. Biplot of logs of finger limens: weights vs. individuals (averaged over rates).

the weights--individuals combinations: Θ_{kw} is a weight-individual effect and β_r, α_r are the slope and intercept for rate r. Note that this model is a little more general than the one we had before, which we could obtain by setting $\beta_r = 1$ for all r in the present model.

Since the form of the individuals-weights interaction is still uncertain, we consider one more biplot. Figure 15 shows the individuals-weights responses averaged over the four rates.

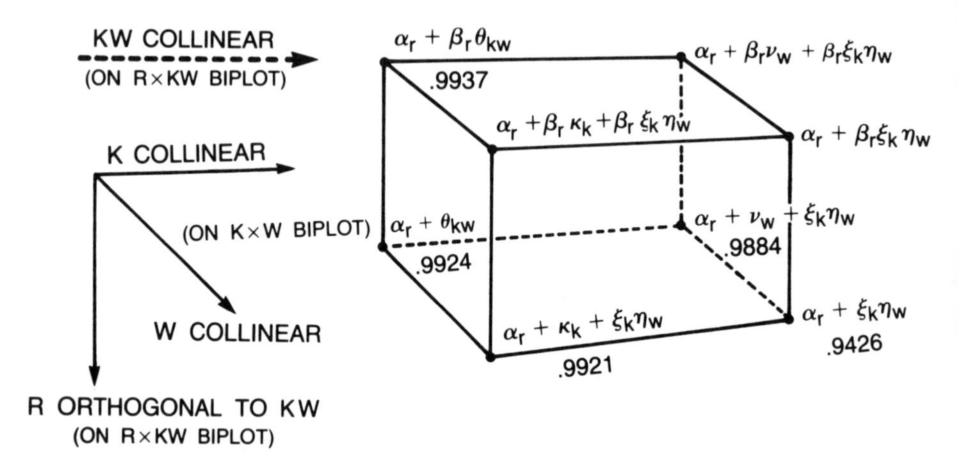

Fig. 16. A summary of models for logs of finger limens.

Here again K (a male) is seen to be very different from L and M (females) and the weights show nice collinearity. The second row of the diagnostic table (Figure 7) applies, so the model for Θ_{kw} is $\Theta_{kw} = \kappa_k + \xi_k\eta_w$, a regression of individuals onto weights.

All these models can be pulled together schematically as shown in Figure 16. Each vertex of the cube is identified with a different model, ranging from the most general (and best fitting) at the top left back corner, to the most specific (and least well fitting) at the bottom right front corner. The biplot diagnostics which indicate specific changes in the model are indicated as directions around the cube. Thus, biplot collinearity of KW-markers suggested the general model $\alpha_r + \beta_r\Theta_{kw}$. Orthogonality of the R-markers to the KW-markers diagnosed absence of R vs. (KW) interaction: the downward arrow therefore indicates modelling in which the R effects are additive to (KW) effects, e.g. $\alpha_r + \beta_r\Theta_{kw}$ becomes $\alpha_r + \Theta_{kw}$.

Collinearity of the K-markers (on the K x W biplot) diagnoses a regression of W onto K: the rightward arrow indicates models in which K effects appear only as "regressors", e.g., Θ_{kw} becomes $\nu_w + \xi_k \eta_w$ and $\kappa_k + \xi_k \eta_w$ becomes $\xi_k \eta_w$. Similarly, collinearity of the W-markers (on the K x W biplot) diagnoses regression of K onto W: the forward arrow thus indicates that W effects appear only as "regressors".

The original fit of the most general $\alpha_r + \beta_r \Theta_{kw}$ model, diagnosed by KW-collinearity, was 0.9937. Additional diagnoses make for more specific models which are more easily interpretable but fit less well. Thus, the diagnosis by orthogonality simplifies the model to $\alpha_r + \Theta_{kw}$ while hardly worsening the fit. On the other hand, the K collinearity diagnosis appreciably reduces the fit in this example. A good model to settle on might be $\alpha_r + \kappa_k + \xi_k \eta_w$ which has a goodness-of-fit of 0.9921--this was diagnosed by the W-collinearity alone.

Again, biplot patterns have diagnosed models very similar to those suggested by John Tukey (1977) using pencil-and-paper EDA methods.

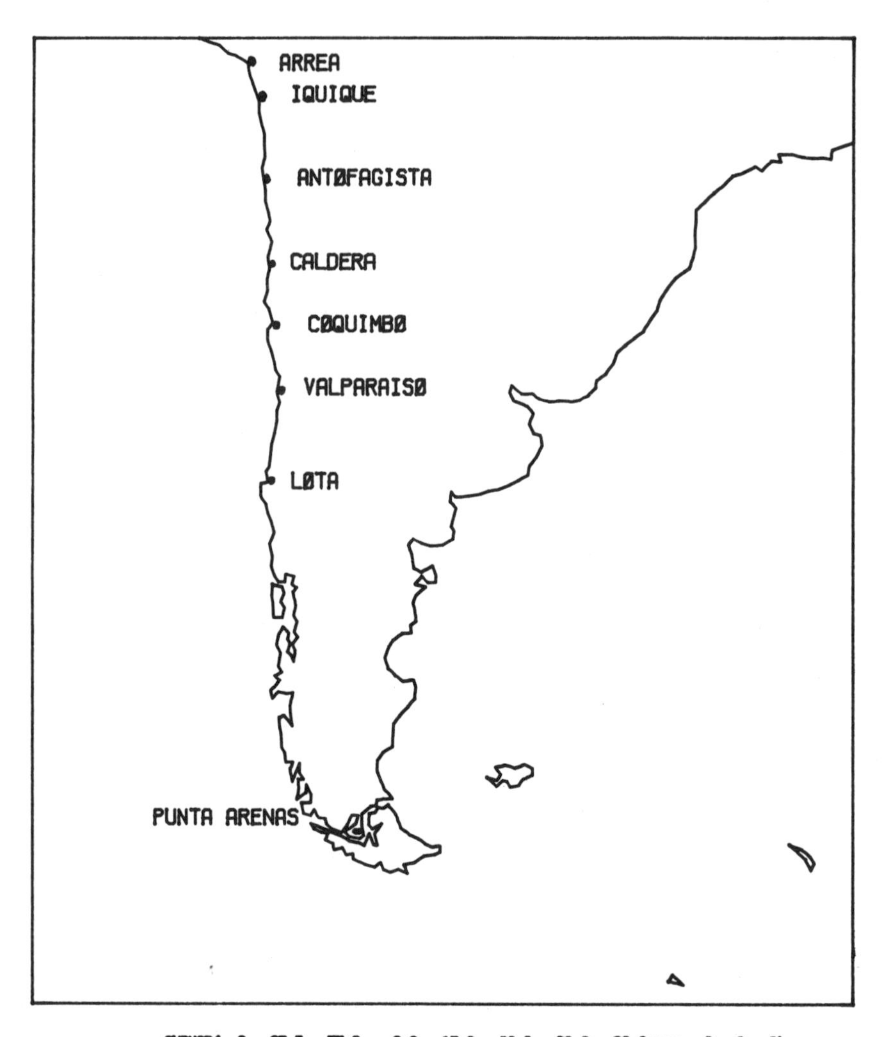

Fig. 17. Seven South American ports.

Our last example deals with ports along the western coast of South America--Figure 17. Ship route distances between these ports are given in Table V A. To begin with, these are arranged in a North-South order so they are a little easier to look at--Table V B. Then the mean distance is subtracted out--Table V C. This makes for a strange kind of

TABLE V A. Shiproute Distances Between S. American Ports.[a]

(distances in sea miles)

	Ant	Arr	Cal	Coq	Iqu	Lota	P A	Val
Ant	0	325	215	396	224	828	1996	576
Arr	325	0	522	702	110	1134	2301	882
Cal	215	522	0	196	420	628	1795	376
Coq	396	702	196	0	602	455	1623	203
Iqu	224	110	420	602	0	1033	2201	782
Lota	828	1134	628	455	1033	0	1191	268
P A	1996	2301	1795	1623	2201	1191	0	1432
Val	576	882	376	203	782	268	1432	0

[a]From J. W. Tukey, EDA, Chapter 11; and the 1963 World Almanac and Book of Facts, New York: World-Telegram and Sun.

TABLE V B. N-S Order.

	Arr	Iqu	Ant	Cal	Coq	Val	Lota	P A
Arr	0	110	325	522	702	882	1134	2301
Iqu	110	0	224	420	602	782	1033	2201
Ant	325	224	0	215	396	576	828	1996
Cal	522	420	215	0	196	376	628	1795
Coq	702	602	396	196	0	203	455	1623
Val	882	782	576	376	203	0	268	1432
Lota	1134	1033	828	628	455	268	0	1191
P A	2301	2201	1996	1795	1623	1432	1191	0

TABLE V C. Mean-Centered Data.

	Ant	Arr	Cal	Coq	Iqu	Lota	P A	Val
Ant	-732.	-407.	-517.	-336.	-508.	96.	1264.	-156.
Arr	-407.	-732.	-210.	- 30.	-622.	402.	1569.	150.
Cal	-517.	-210.	-732.	-536.	-312.	-104.	1063.	-356.
Coq	-336.	- 30.	-536.	-732.	-130.	-277.	891.	-529.
Iqu	-508.	-622.	-312.	-130.	-732.	301.	1469.	50.
Lota	96.	402.	-104.	-277.	301.	-732.	459.	-464.
P A	1264.	1569.	1063.	891.	1469.	459.	-732.	700.
Val	-156.	150.	-356.	-529.	50.	-464.	700.	-732.

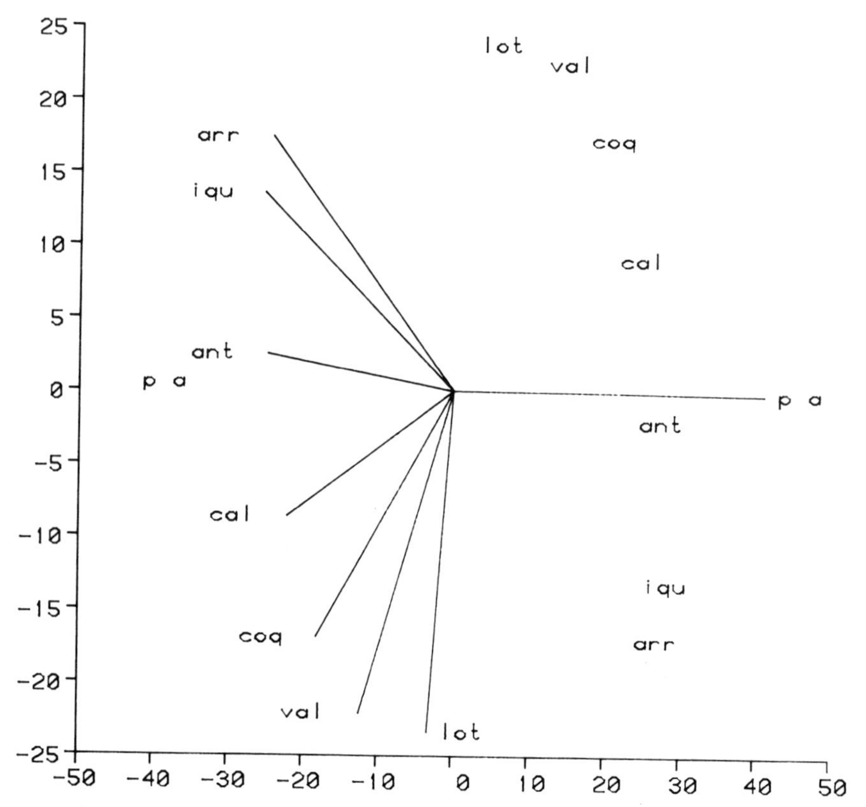

Fig. 18. Biplot of distances between South American ports (mean-centered).

distance, but one that is easier to biplot. Figure 18 shows the biplot of these mean-centered distances. It is immediately evident that the row markers are an exact reflection of the column markers. This is not really surprising since the matrix is symmetric.

It is of interest to consider what is special about biplots of symmetric matrices. Since $Y = Y'$ it follows that in the factorizations $AB' = BA'$. Thus, one may wonder whether $A = B$ or $A = -B$, or what else may account for this symmetry. If $A = B$, one may display an ordinary biplot of factorization AA' in which row markers coincide with column markers: one set of markers suffices. On the other hand, if A = -B, the display of factorization -AA' leads to a biplot like Figure 18 in which the row markers are reflections $-\underline{a}_i$ of the corresponding column markers $+\underline{a}_i$. This redundancy on the biplot can be eliminiated by displaying markers $\underline{a}_i$ along imaginary axes--an imaginary biplot, so to say. One may also achieve this by displaying only the $\underline{a}_i$'s--and not displaying their negatives--but defining the representation by means of the negative inner product, i.e., $y_{i,e} = -\underline{a}_i'\underline{a}_e$. Geometrically, this can be visualized exactly as an ordinary inner-product except that the sign is negative when the $\underline{a}_e$ projection onto $\underline{a}_i$ is in the direction of $\underline{a}_i$ and positive if it is in the opposite direction. That is quite easy to use in practice. Algebraically, we should be thinking of factorization $\Delta\Delta'$ where the e-th row of Δ is $\underline{\delta}_e' = i\underline{a}_e'$ so that $\underline{\delta}_e'\underline{\delta}_g = (i\underline{a}_e)'(i\underline{a}_g) = -\underline{a}_e'\underline{a}_g$. And the representation of $\underline{\delta}$'s along two imaginary axes looks exactly like that of the $\underline{a}$'s but the imaginary units on the axes produce negative inner-products. (See also Gabriel, 1978, for a biplot with one real and one imaginary axis.)

In the present example of distances the representations and inner-product relations are shown in Figure 19. Large distances, small distances and average distances translate to mean-centered distances above zero, below zero and about zero, respectively.

Distance e to g	Mean-Centered Distance $y_{e,g}$	$\underline{a}_e'\underline{b}_g = \underline{\delta}_e'\underline{\delta}_g$	$\underline{a}_e'\underline{a}_g$	If a's lengths constant: angle $(\underline{a}_e,\underline{a}_g)$	If a's lengths constant: $\underline{a}_e$, $\underline{a}_g$
Large	>0	>0	<0	$(\pi/2,\pi]$	distant
Average	0	0	0	$\pi/2$	orthogonal
Small	<0	<0	>0	$[0,\pi/2)$	close

Fig. 19: On the biplot representation of geographic distances.

In the ordinary biplot representation these $(y_{e,g} - \bar{y})$'s are properly represented by $\underline{a}_e'\underline{b}_g$'s. But that is equal to $\underline{\delta}_e'\underline{\delta}_g = (i\underline{a}_e)'(i\underline{a}_g)$. If one leaves out the i and keeps only the real part, then the sign changes, i.e., $\underline{a}_e'\underline{a}_g = -\underline{\delta}_e'\underline{\delta}_g$. Thus for large distances $y_{e,g}$, the inner-product $\underline{a}_e'\underline{a}_g$ would be negative; for small distances, it would be positive; and for average distances, it would be about zero. Moreover, if the $\underline{a}$'s are all of equal length, the inner-product is simply the cosine and varies as the distance between the $\underline{a}$ points. What this means is that when $\underline{a}_e'\underline{a}_g$ is large, there is an obtuse angle. When $\underline{a}_e'\underline{a}_g$ is zero, there will be a 90° angle, and when $\underline{a}_e'\underline{a}_g$ is positive, the angle will be acute. In terms of distances between the $\underline{a}$'s these correspond to large, average and small distances, respectively. This relation between the $\underline{a}$'s thus turns out to be the same as the relation of the original distances $y_{e,g}$. That is why it was convenient to mean-center these distances: the representation along two imaginary axes turned out to be much the same as the original pattern of distances. With this in mind, it is enough to plot the $\underline{a}$'s, which is equivalent to plotting the $\underline{\delta}$'s along imaginary axes, and to consider only the column markers on Figure 18.

This example shows that on occasion one can make use of imaginary biplots for good display of data. John Tukey's (1977) treatment was quite different. Instead of looking at the data, he first tried a model which was intuitively appealing. He postulated that the distance between port e and port g is the sum of (1) a local distance l_e from port e to the shipping lane, (2) a distance $p_{e,g}$ along the shipping lane, and (3) a distance l_g from the lane into port g. He further postulated that shipping lane distances $p_{e,g}$ are simply additive, thus $p_{1,4} = p_{1,2} + p_{2,3} + p_{3,4}$, etc. This model is shown in Figure 20. If one takes

Distance	Port 1	Port 2	Port 3
Port 1	0	$l_1 + p_{1,2} + l_2$	$l_1 + p_{1,2} + p_{2,3} + l_3$
Port 2	(Fill	0	$l_2 + p_{2,3} + l_3$
Port 3	by symmetry)		0

Fig. 20. Tukey's model for nautical distances.

any tetrad on one side of the diagonal of this distance matrix, its four points show additivity, e.g., $(l_1 + p_{1,3} + l_3) - (l_1 + p_{1,5} + l_5) - (l_2 + p_{2,3} + l_3) + (l_2 + p_{2,5} + l_5) = 0$. On the other hand, a tetrad across the diagonal does not have zero differences. In other words, Tukey's model has $y_{i,j} - y_{i,g} - y_{e,j} + y_{e,g} = 0$ whenever $i < e < j < g$. What would happen in factorization $Y = AA'$ (or $Y = \Delta\Delta'$) of such a matrix? The above tetrad condition is readily seen to become $(\underline{a}_i - \underline{a}_e)'(\underline{a}_j - \underline{a}_g) = 0$ for $i < e < j < g$. In other words, $\underline{a}_i - \underline{a}_e$ is orthogonal to $\underline{a}_j - \underline{a}_g$ whenever $i < e < j < g$. A display of such a model for eight ports is readily seen to require eight vectors $\underline{a}_1, \ldots, \underline{a}_8$ such that $\underline{a}_1 - \underline{a}_2$, $\underline{a}_3 - \underline{a}_4$, $\underline{a}_5 - \underline{a}_6$ and $\underline{a}_7 - \underline{a}_8$ are mutually orthogonal. These are only part of the orthogonalities postulated by the model, but they already require a seven dimensional space to represent them. Evidently such a model cannot be diagnosed on a biplot which is two-dimensional, nor on a 3D bimodel. It is essentially a higher dimensional model.

We have discussed this model in some detail because it is an example of what a biplot cannot diagnose. We have found the biplot to be good for diagnosing some models which are (close) to being two or three dimensional, but this is a case of a model which the biplot just cannot represent because the model cannot be collapsed into a plane or three-space.

* * * *

Finally, what have we learned from these examples? How does biplot inspection compare with the EDA methods proposed in Tukey's (1977) book? Parenthetically, we want to remark that the issue is not one of pencil-and-paper methods of median polish versus computer fitting by least squares, because EDA methods have been computerized. The issue we are addressing is which method gives more insight into the form of models that fit the data. Our experience suggests that in using the biplot, a few displays suffice to reveal relevant patterns in a pretty striking manner. EDA, on the other hand, requires several stages of median polish, inspection of residuals, modelling, and re-expression, further median polish, etc., until one may diagnose a model. The biplot is more immediate: It allows one to see things at a glance.

EDA may show more detail if one inspects fits and residuals carefully at each stage, but it requires iterative cycles of modelling, fitting, residuals inspection, re-expression and decisions. If one fit is inadequate, another is tried until a model is judged adequate. This is a search by trial and error rather than by a systematic method. Moreover, the decisions on model choice are based at each stage on I.I.I.--inspired inspection of irregularities. Irregularities are provided by data, inspection takes time, but inspiration is something that may be difficult to come by. In summary, the EDA modelling procedure is in general not systematic. (An exception to this is Tukey's diagnostic plot which uses comparison values systematically for diagnosis of models and re-expressions.)

Biplot diagnostics are more systematic and direct. One does not start by guessing a model, but rather displays data and inspects it--the diagnosis is then often immediate. We know what biplot collinearities mean; we know what right angles mean; we know what coplanarity means and we know something about distances. Identification of any of these patterns makes modelling automatic and hence, to a large extent, objective. And yet, there is also an interactive and somewhat subjective aspect to biplot modelling. One may use one's prior knowledge about the subject matter to choose among various patterns apparent on the biplot. Tsianco (1980) and Gabriel saw the ellipse in the temperature data, though they were not looking for it and at that time had no idea of how to use such a pattern. But as they traced the seasonal variation of the monthly temperatures, they were led to the elliptical pattern. Similarly, when we identify subtables with simple patterns, we interact with the data's display. So biplot modelling is partly systematized and yet allows the investigator to interact with his data and look for interesting patterns.

To sum up, we have sought to demonstrate, by the examples of this paper, that the EDA methods presented in Tukey's (1977) "Golden Book" are not the only ones available. Much can also be learned about suitable models, and most of the messy trial and error of EDA can be avoided, by displaying the data in a biplot.

REFERENCES

Bradu, D. and Gabriel, K. R. (1978). "The Biplot as a Diagnostic Tool for Models of Two-Way Tables," Technometrics, 20, 47-68.

Cox. C., Davis, H. T., Wardell, W. M., Calimlim, J. F., and Lasagna, L. (1980). "Use of the Biplot for Graphical Display and Analysis of Multivariate Pain Data in Clinical Analgesic Trials," Submitted to Controlled Clinical Trials.

Gabriel, K. R. (1971). "The Biplot Graphic Display of Matrices with Application to Principal Component Analysis," Biometrika, 58, 453-467.

Gabriel, K. R. (1978). "The Complex Correlational Biplot," Theory Construction and Data Analysis in the Social Sciences (S. Shye, ed.) San Francisco: Jossey-Bass, 350-370.

Gabriel, K. R. (1980). "Biplot Display of Multivariate Matrices for Inspection of Data and Diagnoses," Interpreting Multivariate Data. (V. Barnett, ed.) London: Wiley (To appear).

Gabriel, K. R., Rave, G. and Weber, E. (1976). "Graphische Darstelling von Matrizen durch das Biplot," EDV in Medizin und Biologie, 7, No. 1, 1-15.

Gabriel, K. R., and Zamir, S. (1979). "Lower Rank Approximation of Matrices by Least Squares with any Choice of Weights," Technometrics, 21, 489-498.

Haber, M. (1975). "The Singular Value Decomposition of Random Matrices," Ph. D. thesis at Hebrew University, Jerusalem.

Householder, A. S., and Young, G. (1938). "Matrix Approximation and Latent Roots," Am. Math. Monthly, 45, 165-171.

Kester, Nancy K. (1979). "Diagnosing and Fitting Concurrent and Related Models for Two-Way and Higher-Way Layouts," Ph. D. thesis at University of Rochester, Rochester, New York.

McNeil, D. R., and Tukey, J. W. (1975). "Higher-Order diagnosis of Two-Way Tables," Biometrics, 31, 487-510.

Mosteller, F., and Tukey, J. W. (1977). Data Analysis and Regression, Reading, Massachusetts., Addison-Wesley.

Tsianco, M. C. (1980). "Use of Biplots and 3D-Bimodels in Diagnosing Models for Two-Way Tables," Ph. D. thesis at University of Rochester, Rochester, New York.

Tukey, J. W. (1977). Exploratory Data Analysis, Reading, Massachusetts., Addison-Wesley.

THE USE OF SMELTING IN GUIDING RE-EXPRESSION

John W. Tukey

Research-Communications Principles Division
Bell Laboratories
Murray Hill, New Jersey
and
Department of Statistics
Princeton University
Princeton, New Jersey

I. INTRODUCTION

Since guiding and even re-expression are not as familiar as they should be, smelting will come only at the end of this account. The order of business will be:

- Some cases in which one would like to guide re-expression.
- How $\ln|\Delta v/\Delta u|$ can be used for guidance.
- How smelting can be used to get usable values of $\ln|\Delta v/\Delta u|$.

After this, we will work a numerical example, and learn some further lessons from it.

II. GUIDANCE DESIRED

Our discussion of desired guidance will first (Section A) discuss the improvement of regression by re-expression, either of the response (y) or of one or more carriers (x's), pointing out the advantage of extracting naturally smooth parts before applying a nonlinear smoother. Here the aim is a more complete description (explanation?) by the regression. Then we will turn to the question of symmetrizing batches (Section B), pointing out the advantages and convenience of median rankits (order statistic working values). Here the aim is to preserve the local structure of the values while making the large-scale structure nicer. With several batches (Section C) we may, but need not, also strive to bring them to similar variability.)

Our discussion (Section D) of following guidance to find a functional form (a fuform) applies to all the kinds of guidance just mentioned. It emphasizes matching the logarithms of divided differences based on a smooth version of the data to logarithms of derivatives of simple functional forms: trivially re-expressed monomials ("trems") and trivially re-expressed exponentials ("trexs").

Prepared in part in connection with research at Princeton University sponsored by the Army Research Office (Durham).

ISBN 0-12-438180-4

Our discussion of extracting a smooth -- usually monotone -- version of the observed rough relation of a basis of guidance to a variate to be re-expressed proceeds by stages. First (Section E) we (i) replace vertical ties by single values, then (ii) do a robust-resistant smooth, and finally (iii) replace horizonrtal ties bv single points. Then (Section F) we use smelting (smoothing-guided excision) to make the divided differences based on the remaining points smoother -- indeed, often monotone.

Our discussion of an example, drawn from Daniel and Wood (1972, 1980), opens with smoothing and smelting the data (Section G) and proceeds to the choice of functional form, first try in Section H, better tries in Section I, comparative choice in Section J. Pictures of certain stages are shown and discussed in Section K.

Finally, Section L summarizes our conclusions.

A. Regression to be performed

Suppose first that we have data $(y; x_1, x_2, \ldots, x_k)$ and have, at least as a beginning, fitted y with $\hat{y} = a + b_1x_1 + b_2x_2 + \ldots b_kx_k$. What might we want to do so that a fit of the same form -- in which one or more variables has been re-expressed -- will fit better? The simplest things are to

- replace x_1 by some $\phi_1(x_1)$, or
- replace x_2 by some $\phi_2(x_2)$, or
- replace y by some $\phi^{-1}(y)$, or
- make two or more of such replacements (also for $x_3, x_4, \ldots, x_k$ of course).

In doing any of these things, we would like to *guide* the choice of our re-expression. If the given y's are "smooth" enough, we naturally think of:

- guiding $\phi_1(x_1)$ by $(y - \hat{y}) + b_1x_1 = y - (\hat{y} - b_1x_1)$
- guiding $\phi_2(x_2)$ by $(y - \hat{y}) + b_2x_2 = y - (\hat{y} - b_2x_2)$
- guiding $\phi(\hat{y})$ by $y = (y - \hat{y}) + \hat{y}$

all of which have a common form, namely

guiding the re-expression by "residual + already"

where "already" is what was already taken care of by the carrier (or fitted response) that is to be replaced.

What if the y's are not as smooth? Clearly we should smooth something. Does it matter what? Yes, because good smoothers are *non*linear.

* choice of smoother *

With nonlinear smoothers it pays to smooth as little of the variation as possible -- this means leaving out -- not trying to smooth again -- whatever varies much and is already smooth. Since "already" is a smooth function of the carrier (or fitted response) that concerns us, this means that we should smooth the "residual" and leave the "already" alone.

We can now see one reason for finding $\phi^{-1}(y)$ rather than $\phi(y)$, since smoothing $y - \hat{y}$ against y would merely focus on the idiosyncrasies of y itself. We dare not do such a thing!

Let us write

$$(y-\hat{y})_{sm}$$

for whichever one of the several possible smooths of our residuals that is relevant in a particular case. Our basic guider is now modified to be

smoothed residual + already

which calls, more specifically, for:

- guiding $\phi_1(x_1)$ by $(y-\hat{y})_{sm} + b_1x_1$
- guiding $\phi_2(x_2)$ by $(y-\hat{y})_{sm} + b_2x_2$
- guiding $\phi^{-1}(\hat{y})$ by $(y-\hat{y})_{sm} + \hat{y}$

where $(y-\hat{y})_{sm}$, $(y-\hat{y})_{sm}$, and $(y-\hat{y})_{sm}$ differ -- because they are smoothed against x_1, against x_2, or against $\hat{y}$, respectively.

B. Batch and rankits

Suppose we have a batch (shortly more than one!) of values $y_1, y_2, \ldots, y_n$ — which we may as well suppose ordered — and a distribution shape which we would like $\phi(y)$ to follow at least roughly. (In principle, possibly the unit Gaussian; in practice, probably the unit logistic.)

* preliminary groundwork *

It is rare that the raw values of y can be only trusted to tell us about order. If, for instance, we have

label	y-value
A	1237
B	1241
C	1329

we usually have no doubt that we want $\phi(y)$ for A and $\phi(y)$ for B much closer to each other than $\phi(y)$ for B and $\phi(y)$ for C are going to be. (This is, in part, because A, B, and C appear to be moderately near one another -- because they are about the same size. We would not be as sure for 7, 11, and 129.)

If all the y's were close enough together, and $\phi'(y)$ were continuous enough, this would happen.

More generally, it would happen if the *proportionate* change in

$$\frac{d\phi(y)}{dy}$$

were small -- or if this quantity changed only slowly.) (We may need to have the total change large, but we ought to let this only change slowly. That is, of course, if

$$\ln \frac{d\phi(y)}{dy}$$

changed only slowly. Now we can assess this last quantity well enough from

$$\ln \frac{\Delta\phi(y)}{\Delta y}$$

which we now want to at least change slowly. Thus this latter quantity is a natural one to focus on.

* rankits *

Now that we know more about how to preserve in $\phi(y)$ the local properties of y that we would like to preserve -- namely by having $\ln \Delta\phi(y)/\Delta y$ change slowly -- we are ready to bring in the desired approximate distribution shape.

We are working with order statistics $y_1 \leq y_2 \leq \ldots \leq y_n$. We want $\phi(y)$, namely $\phi(y_1) \leq \phi(y_2) \ldots \leq \phi(y_n)$ to resemble, *in the large*, what the order statistics from the given distribution, say with cumulative $F(y)$, *ought* to look like.

And what *ought* the order statistics of a sample of n from the distribution with cumulative $F(z)$ look like? Presumably like some typical values for the respective order statistics -- means perhaps, but medians probably better.

For practical purposes, these medians are the solutions of

$$F(a_{i|n}) = \frac{3i - 1}{3n + 1}$$

The solutions $a_{i|n}$ we will call rankits -- or, more properly, median rankits.

Thus our problem has become:

use the $a_{i|n}$ to guide the choice of $\phi(y_i)$

where we expect

- the guidance to involve choosing

$$\ln \frac{\Delta\phi(y_i)}{\Delta y_i}$$

to follow

$$\ln \frac{\Delta a_{i|n}}{\Delta y_i}$$

- need for care in finding some way to make the latter smooth enough.

C. Several batches

If we ask what guiding by rankits has accomplished for a single batch, we find, in order:

- making the width about right
- making the symmetry roughly right
- doing something about the shape.

Sometimes we would like all three -- sometimes we do not care about controlling width.

For a single batch it does not matter much whether we want to match widths, but for two or more it does. Two distinct problems arise.

Suppose that we have *several* batches $\{x_i\},\{y_i\},\ldots,\{z_i\}$, corresponding sets of rankits $\{a_{i|nx}\}$, $\{a_{i|ny}\}$, . . . , $\{a_{i|nz}\}$, where "nx", "ny"..."nz" are the batch sizes (which may be the same or different) and that we wish to find *one* function $\phi(\ )$ such that

- $a_{i|nx}$ has guided $\{\phi(x_i)\}$
- $a_{i|ny}$ has guided $\{\phi(y_i)\}$
- $a_{i|nz}$ has guided $\{\phi(z_i)\}$

either (1) for all aspects or (2) for all except width.

This means that

$$\ln \frac{\Delta\ new}{\Delta\ guide} = \ln \frac{\Delta\ new}{\Delta\ old} - \ln \frac{\Delta\ guide}{\Delta\ old}$$

is to be nearly constant (in case (2) only within each batch separately). So the problem is to tabulate

$$\ln \frac{\Delta\ guide}{\Delta\ old}$$

after smoothing within each batch, and fit the result with

$$\ln \frac{\Delta\ new}{\Delta\ old}$$

or, alternatively, with

$$\ln \frac{\Delta\ new}{\Delta\ old} + constant$$

where the constant depends only on the batch.

What we need, since $\ln \Delta a_{i|n}/\Delta\phi = -\ln \Delta\phi/\Delta a_{i|n}$

- $\ln \dfrac{\Delta\phi(x_i)}{\Delta a_{i|nx}}$ is roughly constant
- $\ln \dfrac{\Delta\phi(y_i)}{\Delta a_{i|ny}}$ is roughly constant
- $\ln \dfrac{\Delta\phi(z_i)}{\Delta a_{i|nz}}$ is roughly constant

and either (1) these constants are almost the same, when they may as well be unity, or (2) these constants are allowed to be different.

To get the constraints about the same, we need, in case 1, to look at:

- $\ln \dfrac{\Delta a_{i|nx}}{\Delta x_i}$ as a function of x
- $\ln \dfrac{\Delta a_{i|ny}}{\Delta y_i}$ as a function of y
- $\ln \dfrac{\Delta a_{i|nz}}{\Delta z_i}$ as a function of z

and find, somehow, a single function, $\phi(\ )$, such that

- $\ln \dfrac{\Delta\phi(u)}{\Delta u}$ as a function of u

is close to *all* of them.

If our concern is only with shape--width does not matter -- we are entitled to subtract one constant from the

$$\ln \frac{\Delta a_{i|nx}}{\Delta x_i}$$

another constant from the

$$\ln \frac{\Delta a_{i|ny}}{\Delta y_i}$$

and so on.

Now it is enough to fit these adjusted values with

$$\ln \frac{\Delta\phi(u)}{\Delta u}$$

III. GETTING GUIDANCE FROM VALUES

D. Using $\ln(\Delta v/\Delta u)$ to find a fuform (functional form)

All the guidance problems we have discussed come down to:

- getting a good smooth set of values of $\ln(\Delta v/\Delta u)$ from the data (to be discussed below)
- matching these values with those of $\ln(\Delta\phi(u)/\Delta u)$ for some reasonably simple fuform (discussed in this section)

and going on, in the regression case at least, to:

- refitting the coefficients in the fuform -- and then, perhaps, repeating the whole process, either for the same carrier (or fitted response) or another, in search of an improved fit (standard, and hopefully familiar).

If the values of u, the variable to be re-expressed, are all positive:

- we are free to work with $\ln u$
- we are free to consider u^p for general p

If we can fit $\ln(\Delta v/\Delta u)$ with $A + B \ln u$, we want

$$\ln \frac{\Delta\phi(u)}{\Delta u} \approx A + B \ln u$$

which is nearly enough

$$\ln \frac{d\phi(u)}{du} = A + B \ln u$$

which integrates to

$$\phi(u) = C + \left(\frac{e^A}{B+1}\right) u^{B+1}$$

a trivially re-expressed monomial ("trem").

For any u, whether or not always positive, we can try

$$\ln \frac{\Delta\phi(u)}{\Delta u} \approx D + Eu$$

which is nearly enough

$$\ln \frac{d\phi(u)}{du} = D + Eu$$

which integrates to

$$\phi(u) = F + \left(\frac{e^D}{E}\right) e^{Eu}$$

a trivially re-expressed exponential ("trex").

What if we cannot fit with a straight line?

Clearly we can try:

- fitting a broken straight line (when we get a trem-spline or a trex-spline)
- fitting a simple curve (if we can "see" which one to fit), when we may have to use quadrature to define $\phi(u)$
- treating the relation of $\phi(u)$ to u in just the same way as we initially treated the relation of v to u, thus starting the process over (and leaving us with two integrations to be performed, algebraically or numerically).

It is natural to ask "In one of the regression cases we needed $\phi^{-1}(\ )$, are we all right?"

Certainly, since

$$v = \phi_m(u) = C + \left(\frac{e^A}{B+1}\right)u^{B+1}$$

corresponds to

$$u = \phi_m^{-1}(v) = \left(\frac{B+1}{e^A}\right)^{\frac{1}{B+1}}(v-C)^{\frac{1}{B+1}}$$

while

$$v = \phi_x(u) = F + \left(\frac{e^D}{E}\right)e^{Eu}$$

corresponds to

$$u = \phi_x^{-1}(v) = -\left(\frac{D - \ln E}{E}\right) + \frac{1}{E}\ln(v-F)$$

IV. GETTING USABLE VALUES

E. And Now to Prepare!

The regression problems give us sequences of pairs (u_i,v_i) where v is trying to guide u. These may be far from smooth.

The following steps are usually in order:

- if there are several v_i for a single value of u_i, replace the (u_i,v_i) pairs by one pair $(u,med\{v_i\})$. (We now have a sequence $\{(u_i,v_i)\}$ with $u_1 < u_2 < \cdots$)
- smooth — in a robust-resistant manner — the values of the v_i until they too are monotone — say $v_1 \leq v_2 \leq v_3 \ldots$
- if there are two or more u_i for a single value of v, replace the corresponding pairs (u_i,v_i) by one pair $(med\{u_i\},v_i)$.

When we have done this, we will have a double, strictly monotone paired sequence $\{(u_i,v_i)\}$, which is likely to have fewer points than we began with.

How are we to do the smoothing in the second step? How can we easily force monotonicity?

- we may begin by replacing each v_i by the median of the three adjacent values v_{i-1}, v_i, v_{i+1}. (This is "3")
- then we can repeat this as many times as anything changes. (Together, this is "3R")
- then, we can copy out the v's in order, omitting adjacent repetitions, and repeat the earlier steps
- and then we can repeat the third step as often is necessary
- if need be, we can find the smoothed value for each original v_i by tracing through the sequence at values corresponding to it after each successive third step

This process can only stop when the v's are strictly monotone -- since no equal adjacent pair survives each third step.

F. And now for smelting!

The guiding for batches (from rankits) makes a monotone sequence: the smoothing of the last section makes a monotone sequence. This is enough to make $\Delta v/\Delta u$ nonzero and of a common sign. As a result it makes sense to look at one or the other of

$$\ln\left(\frac{\Delta v}{\Delta u}\right) \quad \text{and} \quad \ln\left(-\frac{\Delta v}{\Delta u}\right)$$

one of which will be real in all cases. (It may of course be of either sign or both signs.) We are now ready to use a less familiar kind of smoothing, *smoothing by excision* (rather than smoothing by value change). If we *excise* some of the (u_i,v_i) pairs, renumber those that are left and find the new values of $\ln(\Delta v_i/\Delta u_i)$, we won't be in any trouble and will have preserved the measuring of the relation of the v's to the u's.

One thing we certainly will want in the vast majority of cases, is to have the values of

$$\ln(\Delta v_i/\Delta u_i) \quad \textit{monotonic}$$

which is the same as having

$$\Delta v_i/\Delta u_i \quad \textit{monotonic}$$

* gaining monotonicity *

Let us look at v against u. One thing that may happen is shown in the adjacent figure.

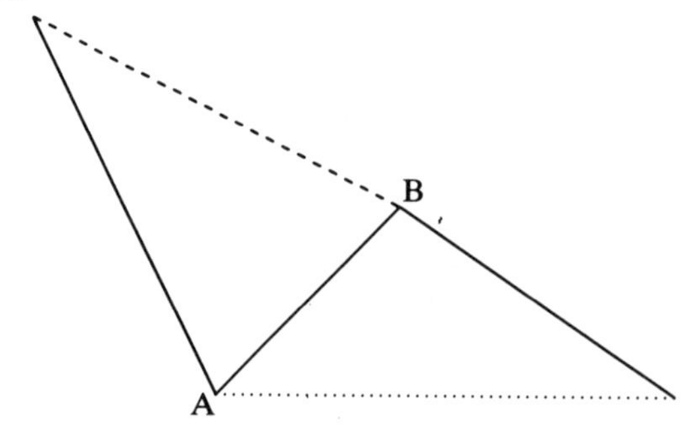

If we excise point A, we get the dashed segment ------ in place of two solid segments.

If we excise point B instead, we get the dotted line segment ...... in place of two solid segments (one the same, the other different).

It seems reasonably natural to combine the center segment with the one nearest it in slope.

One way to do this is to:

- do a 3R smooth on the values of $\ln(\Delta v/\Delta u)$, and
- whenever 2 adjacent values of the smoothed sequence are equal, delete the intervening point.

Notice that

- we will sometimes delete (= excise) two or more points "in a row",
- we can keep repeating the process until there is no change,
- this will ensure a monotone sequence of values for the final $\{\ln(\Delta v/\Delta u)\}$ -- since no equal adjacent pair survives a full step, all local minima or maxima will be simple, and will thus be changed.

Two, somewhat minor, special cases deserve our attention:

- sometimes the 3R smoothing will change $-,-,a,b,-,-$ into $-,-,b,a,-,-$. This, too, is a reason for *excising* the intervening point
- we have provided no way to change or excise the end points — something is badly needed here.

●● a plausible procedure, not yet adequately tested, would go as follows:

●●● look at the signs of

$$\Delta \ln(\Delta v/\Delta u)$$

for each new calculation of same

●●● if either sign has a 2/3 (?(3/4)?) preponderance, and an end sign is opposite, delete (excise) the corresponding end pair -- (u_i,v_i) for one extreme value of i or the other (or conceivably both).

Because it "burns away" the less useful points, while keeping the more useful ones, we call this process *smelting*. (Notice also the occurrence of "sm" in smoothing and smelting.)

V. A WORKED EXAMPLE AND SOME FURTHER LESSONS

G. Smoothing and smelting the example.

We now consider a single worked example, one *not* intended to show all that can happen. (No one example could show all that might happen.)

At the start of their chapter 7, Daniel and Wood (1972, 1980) discuss a 10-variable regression problem in considerable detail. They call attention to their plot of residuals against x_3 (in the form of a component-plus-residual plot) saying (page 125) "...we can see the observations suggest curvature, possibly in the form of a squared term. Excluding observation 1 would make the curvature more pronounced."

We shall work with 100 times D&W's x_3 and 1000 times their residuals in order to avoid unnecessary decimals, calling them just "x_3" and "residual". Exhibit 1 shows the residuals (and serial identification) for ordered values of x_3 and the combination of few/several residuals, for the same x_3, into their median.

exhibit 1

The residuals ordered on x_3 and combined into medians where x_3 is tied

(100x) x_3	serial	(1000x) Residual	Median
42	28	-109	
42	30	-18	-18
42	31	91	
43	25	64	
43	26	35	50
44	27	-169	
44	29	-15	-92
46	32	0	
47	33	19	
47	35	66	42
48	24	15	
49	6	-99	
50	2	-25	
50	7	-68	
50	10	-71	-70
50	34	-146	
51	4	-31	
51	5	107	
51	21	-28	40
51	23	159	
53	11	88	
54	15	52	
54	22	-16	18
55	9	1	
57	1	-218	
59	17	29	
60	13	184	
62	14	16	
62	37	8	16
62	38	88	
63	12	-79	
63	36	-74	-76
65	16	154	
68	3	81	
78	19	6	
84	8	-148	

Exhibit 2 shows a 3RSS smoothing procedure applied to the results of exhibit 1. It also shows the result of combining ties again, this time ties in the smoothed values for which we replace the few/several points by one point at the median x_3.

exhibit 2
3RSS smoothing* applied to
the medians of exhibit 1

x_3	median	3RSS smooth*				smooth
42	-18					-18
43	50	-18				-18
44	-92	0				0
46	0	0				0
47	42	15				15
48	15	15				15
49	-99	-70	15			15
50	-70	-70	40	15		15
51	40	40	-70	40	18	18
53	88	40	40	18		17
54	18					18
55	1	1	1	18		18
57	-218	1	29	1	16	16
59	29	29	1	16		16
60	184	29	16			16
62	16	16	29			29
63	-76	16	81	29		29
65	154	81	16	81	29	29
68	81	81	81	16		16
78	6					6
84**	-148	-148			-14	-14

final ties		results of interchange
42h	-18	
45	0	
48h	15	
53h	18	16
59	16	18
64	29	
68	16	
78	6	
84	-14	

*See *Exploratory Data Analysis,* (Tukey 1977) chapter 7 for definitions and examples of this type of smoothing.

*Unchanged values often omitted.

**The change from -148 to -14 is *not* a typographical error, but rather an application of the end-value rule.

We started with 36 points (#8 and #18 missing). Exhibit 2 reduced this to 20, and exhibit 2 reduced it to 9, an ordinarily useful degree of compaction.

While general smoothing of this reduced set threatens to lose us much of what we want, if interchanges of adjacent values (which cannot alter the values of meaningful maxima and minima) will increase smoothness, we should make them. This is done in one instance in exhibit 2.

We are now ready to readd, forming

$$b_3x_3 + (smoothed\ residual)$$

at each of the 9 points at which a finally smoothed residual is available. Exhibit 3 does this, calculates lodids (logs of |divided differences|), and smooths them. The final panel includes mid values of x_3 and ln x_3 (together with ln *mid* x_3, included to show how little different it is from mid ln x_3).

exhibit 3
Readding, lodid formation, and smoothing
(b_3 = -9.94835 in our units)

A). First round of smelting.

x_3	smooth residual	b_3x_3	readd	lodid*	smooth(3R)**	surviving***
42h	-18	-423	-441			(42h, -441)
				1.030		
45	0	-448	-448			(45, -448)
				1.692		
48h	15	-482	-467			(48h, -467)
				2.282	2.266	
53h	16	-532	-516			
				2.266		
59	18	-587	-569			
				2.054	2.266	
64	28	-637	-608			(64, -608)
				2.565	2.398	
68	16	-676	-660			
				2.398	2.565	
78	6	-776	-770			(78, -770)
				2.592		
84	-14	-836	-850			(84, -850)

B). Second round of smelting.

x_3	readd	lodid	(mid x_3)	(log mid x_3)	(mid log x_3)
42h	-441				
		1.030	(at 43.75)	(at 3.778)	(at 3.778)
45	-448				
		1.692	(at 46.75)	(at 3.895)	(at 3.844)
48h	-461				
		2.208	(at 56.25)	(at 4.030)	(at 4.020)
64	-608				
		2.449	(at 71)	(at 4.263)	(at 4.258)
78	-770				
		2.592	(at 81)	(at 4.394)	(at 4.394)
84	-850				

* ln = $\log_e$ of |divided difference|. (Note that divided differences must have constant sign.)

**unchanged values omitted

***neither (a) between two lodids equal after smoothing (e.g. at 53h), nor (b) between two lodids interchanged by smmothing (e.g. at 68), nor (c) an end value whose adjacent lodid was altered by the end-value rule.

The final smelting table, here the 2nd round, elsewhere perhaps the 3rd, 1st or 6th, contains all the information that we are to allow the functional recognizer to work on. Here it consists of 6 pairs of the form (x, piece of fit PLUS smoothed residual) and 5 pairs of the form (mid x, lodid) or (mid log x, lodid) that come from them. Smoothing and smelting are over. The functional recognizer should not look back beyond these selected points. (Though it will always be possible, once a functional form has been recognized and fitted, to form new residuals and start again, in search of a further correction.)

H. Elementary functional recognition.

We now ask ourselves: Can we fit our lodids linearly in either mid x_3 or mid log x_3? The (first divided) difference tables look like this:

43.75	1.030		3.778	1.030	
		.343			15.83
46.75	1.692		3.844	1.692	
		.054			2.93
56.25	2.208		4.020	2.208	
		.015			1.01
71	2.449		4.258	2.449	
		.014			1.05
81	2.592		4.394	2.592	

Neither one seems contented with a constant slope.

If we trusted our input to the utmost, and wanted only an interpolation formula, we could break the range of x_3 at 56.25 and fit a bearable straight line to each half. But our situation is the reverse: we do not believe in high precision for our input, and we greatly prefer a single functional form for use in regression. So we should look for just that.

If our (mid x, lodid) points had lain along a straight line, we would have found the equation of that line and gone on with our calculation. Given 5 points more or less along 2 lines, it is tempting to think of fitting a line "to the 5 points" and then going on to use this line. Tempting, but not helpful, as we will see.

For simplicity, let us do this by fitting a line to the 2nd and 4th points. The necessary arithmetic, and check, runs as follows:

x_3	lodid	Δ	$.0312(x_3\text{-}60)$	diff
46.75	1.692		-.413	2.105
		.0312		
71	2.449		.343	2.106

so the line is

$$\text{fitted lodid} = 2.106 + .0312(x_3 - 60)$$

which we convert into

$$\ln\left(-\frac{d\hat{y}}{dx}\right) = 2.106 + .0312(x_3 - 60)$$

(recall that our divided differences were all negative). Exponentiating

$$-\frac{d\hat{y}}{dx} = e^{2.106}e^{.0312(x_3-60)} = 8.215e^{.0312(x_3-60)}$$

and integrating

$$\hat{y}-c = -\frac{8.215}{.0312}e^{.0312(x_3-60)}$$

$$= -263.3e^{.0312(x_3-60)}$$

Exhibit 4 now compares the values of $y-c$ thus defined with the readded values we started with. Our residuals are not small, and clearly trend systematically. If this is the best we can do in this way, we should give up a straight line and look for something else.

exhibit 4
The 2nd-and-4th-lodid-point fit completed and assessed

x_3	y	$\hat{y}-c$*	diff	adjusted fit**	residual
42h	-441	-152.5	-288.5	-443	-2
45	-448	-164.9	-283.1	-456	-8
48h	-462	-183.9	-278.1	-475	-13
64	-608	-298.3	-309.7	589	19
78	-770	-461.7	-308.3	-753	17
84	-850	-556.7	-293.3	-848	2
		median	-290.9		0
		(range)	(31.6)		(32)

* $\hat{y}-c$ is -264.4$e^{.0312(x_3-60)}$.
** $\hat{y}-c$ = 290.9

I. Repairing the difficulty linearly.

It is easy to do quite a lot better than this. If we go back to the (x_3, y) points, and select 3 points, we will only have two lodids, which must then lie on a line. If we select the first, last and middle points we have

```
42h  -441
            2.050 (at 53.25)              -.144  2.194
64   -608                      .02135
            2.493 (at 74)                  .299  2.194
84   -850
```

making the line

$$\text{fitted smoothed lodid} = 2.194 + .02135(x_3-60)$$

whence

$$-\frac{d\hat{y}}{dx} = e^{2.194}e^{.02135(x_3-60)}$$

$$\hat{y}-c = 420.2e^{.02135(x_3-60)}$$

which leads to the comparisons in exhibit 5.

exhibit 5
A 3-original-point fit completed and assessed

x_3	y	$\hat{y}-c$*	diff	adjusted fit**	residual	residual***
42h	-441	-289.2	-151.8	-439	2	9
45	-448	-305.1	-142.9	-455	-7	0
48h	-467	-328.7	-138.3	-479	-12	-7
64	-608	-456.7	-151.3	-607	1	-1
78	-770	-617.1	-152.9	-767	3	0
84	-850	-701.4	-148.6	-852	-2	-2
		med	-150.1		0	0
		(range)	(14.6)		(15)	(16)

* $\hat{y}-c = -420.2e^{.02135(x_3-60)}$

** adjusted fit = $\hat{y}-c-150.1$

*** If we keep (45, -448) (64, -608), (84, -850) we get

$$\hat{y}-c = -502.2e^{.01857(x_3-60)}$$

and, eventually, these residuals.

We have, by this simple device, made the residuals considerably smaller and much less systematic. The 3 fits of this kind have |residuals| with

fit	largest	next to largest
2 lodid points	19	17
(42.5,64,84)	12	7
(45,69,84)	9	7

The last of the three is probably about as well as we can do with

lodid fitted by a linear function of x_3

We may as well also look, however, at using (42.5,69,84) and a linear function of $\ln x_3$. The basic calculation is:

mid ln x	lodid	Δ	1.299(ln x-4)	
3.954	2.050		-.060	2.110
		1.299		
4.295	2.493		.383	2.110

with the line

$$\text{fitted lodid} = 2.110 + 1.299(\ln x_3 - 4) = -.3086 + 1.299 \ln x_3$$

whence

$$-\frac{d\hat{y}}{dx} = e^{-3.086} x_3^{1.299} = .04568 x_3^{1.299}$$

and

$$\hat{y} - c = -.01987\, x_3^{2.299}$$

with the completion and comparison in exhibit 6. The residuals are almost as good as those for three-point fits to x_3. (Maximum |residual| is now 11, and next to maximum = 7.)

exhibit 6
A 3-original-point fit to ln x_3 completed and assessed

x_3	y	$\hat{y}-c$*	diff	adjusted fit**	residual
42h	-441	-110.1	-330.9	-434	-7
45	-448	-125.6	-322.4	-460	-12
48h	-462	-149.2	-312.8	-473	-11
64	-608	-282.2	-325.8	-606	2
78	-770	-444.8	-325.2	-769	1
84	-850	-527.3	-322.7	-851	-1
		median	-324.0		
		(range)	(18.1)		(18)

* $\hat{y}-c = -.01987 x_3^{2.299}$

** adjusted fit = $\hat{y} - c$ - 324.0.

J. Which to use?

In the context of Daniel and Wood's example, it would seem that either

$$-420.2e^{.02135(x_3-60)} - 150.1$$

or the result of repeating the least-squares fit with

$$e^{.02135(x_3-60)}$$

in the place of x_3, or

$$-.01987x_3^{2.299} - 324.0$$

or a modification of this using $x_3^{2.3}$, or the result of repeating the least-squares fit with

$$x_3^{2.3}$$

in place of x_3 would be quite fancy enough. (Particularly since Daniel and Wood found a fit using

$$ax_3 + bx_3^2$$

to give no significance for the bx_3^2 term.)

It is to be hoped that this example shows:

- how the arithmetic goes
- that dealing with curved plots can make good use of considerable care.

K. The example in pictures.

Now that we have gone through the arithmetic in our example, it may be helpful to look at the steps pictorially. The elimination of vertical ties is routine enough not to require a picture. Exhibit 7 shows the replacement of the 20 resulting points by 9 smoothed values. Exhibit 8 compares these 9 points with what has been fit to them, namely the difference between the new fit

$$-150.1 - 420.2e^{.02135(x_3-60)}$$

and the original fit

$$-9.94835x_3$$

We see the general satisfactoriness of the results. Exhibit 9 shows the difference of fits against a background of the 20 vertical medians, stressing just how much smoothing has been used. (A picture of the difference in fits against the 36 original points would only make the same point somewhat more vigorously.)

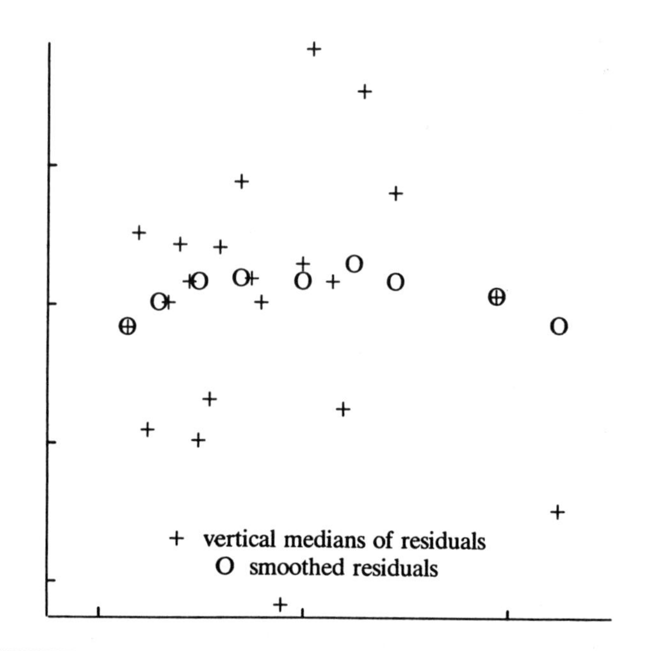

EXHIBIT 7. The 21 vertical medians of residuals and the 9 smoothed residuals

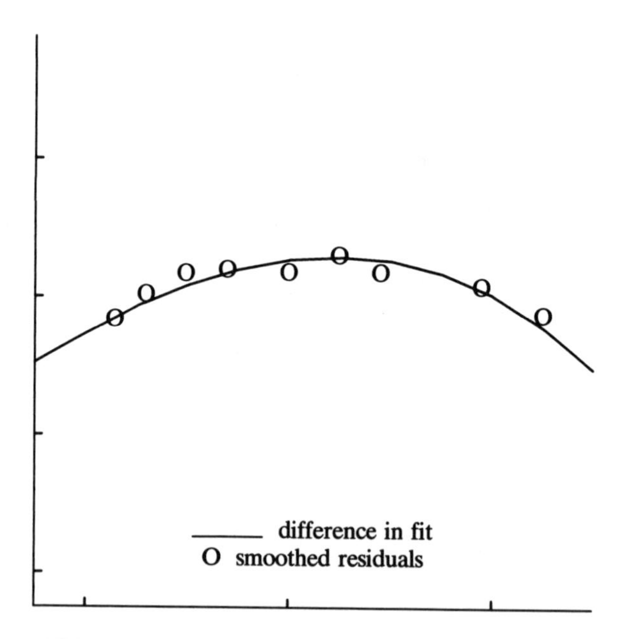

EXHIBIT 8. The 9 smoothed residuals and the difference in fit

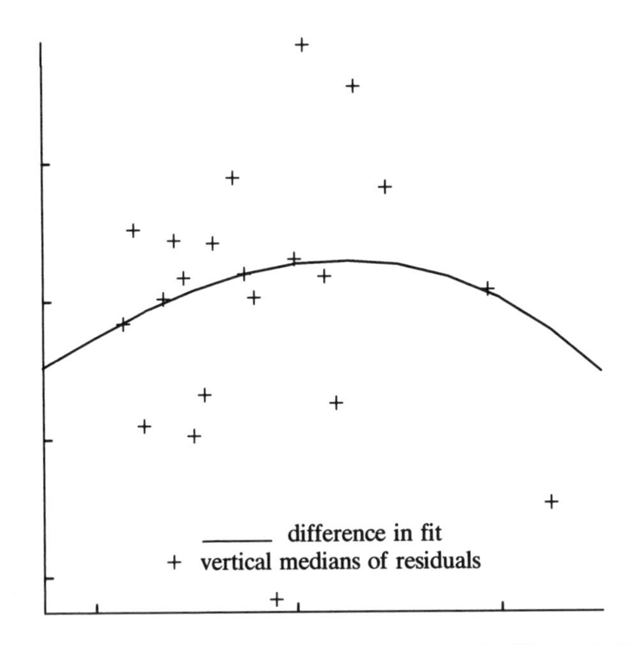

EXHIBIT 9. The 21 vertical medians of residuals and the difference in fit

In assessing these pictures, remember that a good smoothing technique will make anything smooth, and that a good fitting technique will make a fit to anything. The point of this example is not that we have discovered exactly what the dependence on x_3 is like; we are far from doing that. It is that we have effectively found what the dependence seems to be, in view of data that was barely enough to convince Daniel and Wood that there might be curvature.

If we want a fit as an appearance, but surely not a certainty, our technique has worked well. If, for example, we included x_3 to be sure that x_3's effects would not propagate into the other parts of the fit, our new fit is probably the best we know how to make. (This in no sense implies that the true dependence on x_3 is exponential.)

V. SUMMARY

L. Recap

We have shown how either of

- improving a regression fit
- shaping (and sizing) a batch
- shaping several batches simultaneously
- shaping and sizing several batches simultaneously

leads to a situation in which we would like to use one variable to guide the reexpression of another. This leads to values of

$$\ln \frac{\Delta\ guide}{\Delta\ old}$$

which we want to fit with values of

$$\ln \frac{\Delta\ new}{\Delta\ old}$$

thus making small the values of

$$\ln \frac{\Delta\ new}{\Delta\ guide}$$

which corresponds to

$$\Delta\ new \approx \Delta\ guide$$

and can lead us to

$$new \approx guide$$

at least for the points we kept.

In such a process, irregularity can ruin us -- so we avoid it:

- in the regression case by an initial smoothing of the values (guide values) entering into

$$\ln \frac{\Delta\ guide}{\Delta\ old}$$

which suffices to get to monotonicity

- in all cases, after smoothing to monotonicity if need be, by *smelting* the resulting sequences — smoothing them by excision, where the excision is guided by the results of smoothing $\ln(\Delta v/\Delta u)$

(In smelting, the numerical values found by smoothing are *not* used, except for such guidance.)

REFERENCES

Daniel, C. and Wood, F. S. (1972, 1980). *Fitting Equations to Data*. John Wiley, New York.

Tukey, J. W. (1977). *Exploratory Data Analysis*. Addison-Wesley, Reading, MA.

GEOMETRIC DATA ANALYSIS: AN INTERACTIVE GRAPHICS PROGRAM FOR SHAPE COMPARISON

Andrew F. Siegel[1]

Department of Statistics
Princeton University
Princeton, New Jersey

I. INTRODUCTION

The quantitative study of shape and form has been considered by many authors since the fundamental descriptive work of Thompson (1917). The method of least squares was used by Sneath (1967) to solve the problem of finding a common location and orientation of two shapes in order to compare their similarities and differences. A general discussion of the numerical aspects of such least squares calculations may be found in Huber (1980). There are situations in which robust methods such as the repeated median technique (Siegel, 1980) are far superior to least squares, especially in the detection of localized shape differences (Siegel and Benson, 1979).

Least squares and robust methods are both considered here, and an interactive computer program written in FORTRAN with graphics capabilities is presented. Section 2 outlines the mathematical methods involved in each fitting process;

[1]Research supported in part by U.S. Army Research Office Contract DAAG29-79-C-0205.

ISBN 0-12-438180-4

further details can be found in the references. Two examples are provided in Section 3 to illustrate the use of the program and to show the kinds of graphic results that can be obtained. Section 4 outlines the work involved in setting up the program to run on a computer installation and provides an overall description of how to use the system. The interactive commands are described in detail in Section 5, and the program is listed in the Appendix.

II. METHODS

Each of the two shapes is represented by a sequence of n points in two dimensions. Homology is assumed, so that the ith point of shape 1 corresponds to the ith point of shape 2 because, for example, this might be the location of the base of the skull in each specimen. Establishment of homology can be straightforward in some situations, but is difficult in others. We will assume it has been done, even if only tentatively, and will also assume that any needed reflection has also already been done. Thus our data are

Shape 1	Shape 2
(x_1,y_1)	(u_1,v_1)
(x_2,y_2)	(u_2,v_2)
.	.
.	.
.	.
(x_n,y_n)	(u_n,v_n)

Holding shape 1 fixed, we transform shape 2 by a rotation angle θ, a magnification factor ρ, and a translation (a,b). The transformed coordinates of shape 2 are therefore

$$\begin{pmatrix} u_i' \\ v_i' \end{pmatrix} = \begin{pmatrix} a \\ b \end{pmatrix} + \rho \begin{pmatrix} \cos(\theta) & -\sin(\theta) \\ \sin(\theta) & \cos(\theta) \end{pmatrix} \begin{pmatrix} u_i \\ v_i \end{pmatrix} \tag{2.1}$$

Defining $c=\rho\cos(\theta)$ and $d=\rho\sin(\theta)$ we can reparametrize to an expression that is linear in the parameters:

$$\begin{pmatrix} u_i' \\ v_i' \end{pmatrix} = \begin{pmatrix} a \\ b \end{pmatrix} + \begin{pmatrix} c & -d \\ d & c \end{pmatrix} \begin{pmatrix} u_i \\ v_i \end{pmatrix} . \qquad (2.2)$$

The least squares solution can be found by minimizing the sum of squared distances from each point (x_i, y_i) of shape 1 to the transformed point (u_i', v_i') of shape 2. This sum of squares is

$$SS(a,b,c,d) = \sum_{i=1}^{n} [(x_i - u_i')^2 + (y_i - v_i')^2] . \qquad (2.3)$$

Minimizing (2.3) we obtain the least squares parameter values:

$$\hat{c} = \frac{\sum_{i=1}^{n} [(u_i-\bar{u})(x_i-\bar{x}) + (v_i-\bar{v})(y_i-\bar{y})]}{\sum_{i=1}^{n} [(u_i-\bar{u})^2 + (v_i-\bar{v})^2]} \qquad (2.4)$$

$$\hat{d} = \frac{\sum_{i=1}^{n} [(u_i-\bar{u})(y_i-\bar{y}) - (v_i-\bar{v})(x_i-\bar{x})]}{\sum_{i=1}^{n} [(u_i-\bar{u})^2 + (v_i-\bar{v})^2]} \qquad (2.5)$$

$$\hat{a} = \bar{x} - (c\bar{u} - d\,\bar{v}) \qquad (2.6)$$

$$\hat{b} = \bar{y} - (c\bar{v} + d\,\bar{u}) \qquad (2.7)$$

The repeated median procedure estimates the magnification (ρ) and the rotation (θ) separately, using a double median for each. These estimates are

$$\tilde{\rho} = \underset{1\leqslant i\leqslant n}{\text{Median}} \left\{ \underset{\substack{1\leqslant j\leqslant n \\ j\neq i}}{\text{Median}} \; \rho_{ij} \right\} \qquad (2.8)$$

$$\tilde{\theta} = \underset{1\leqslant i\leqslant n}{\text{Median}} \left\{ \underset{\substack{1\leqslant j\leqslant n \\ j\neq i}}{\text{Median}} \; \theta_{ij} \right\} \qquad (2.9)$$

where ρ_{ij} and θ_{ij} are the magnification and rotation parameters computed using only points i and j of each shape. The transformation factors can be estimated using a single median:

$$\tilde{a} = \underset{1 \leqslant i \leqslant n}{\text{Median}} \{x_i - \tilde{\rho}[u_i \cos(\tilde{\theta}) - v_i \sin(\tilde{\theta})]\} \tag{2.10}$$

$$\tilde{b} = \underset{1 \leqslant i \leqslant n}{\text{Median}} \{y_i - \tilde{\rho}[u_i \sin(\tilde{\theta}) + v_i \cos(\tilde{\theta})]\} \tag{2.11}$$

The match obtained using repeated medians, unlike least squares, will not be led astray by a few atypical points and localized changes will be more easily identified. In general, if more than (n+1)/2 (i.e. just over half) of the points can be made to match closely, then the repeated median method will produce a transformation that does match them closely.

III. EXAMPLES

Two examples are given here in order to illustrate the possible results obtainable through the use of this system. The first example is a square (shape 1) compared to a deformed square (shape 2). Using the line drawing capability to connect points 1 to 2, 2 to 3, ..., 7 to 8, and 8 to 1, the initial shapes can be drawn using the commands d1 and d2. This is shown in Figure 1 in which numbers have been added in order to identify homologous points.

Typing the command ℓs causes shape 2 to be transformed to the least squares fit. A display superimposing both shapes can be obtained with the command ds, resulting in Figure 2a. Residual vectors starting from the points of shape 1 and ending at the homologous points of shape 2 can be drawn using the command dr. Figure 2b shows these residuals superimposed on shape 1, using commands dr and d1.

The robust fit by repeated medians is produced by the command rs. Using commands ds, dr, and d1 as before, we can obtain the displays of the fitted shapes and residuals shown in Figure 3 for the robust fit.

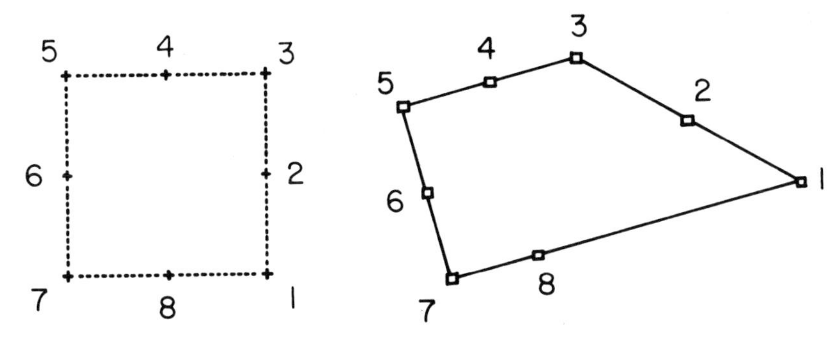

FIGURE 1. Two shapes to be compared.

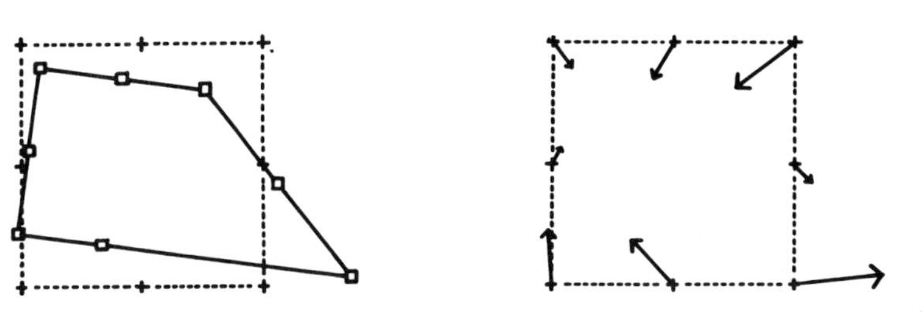

FIGURE 2. Least squares fit. Shapes superimposed (left); residuals superimposed on shape 1 (right).

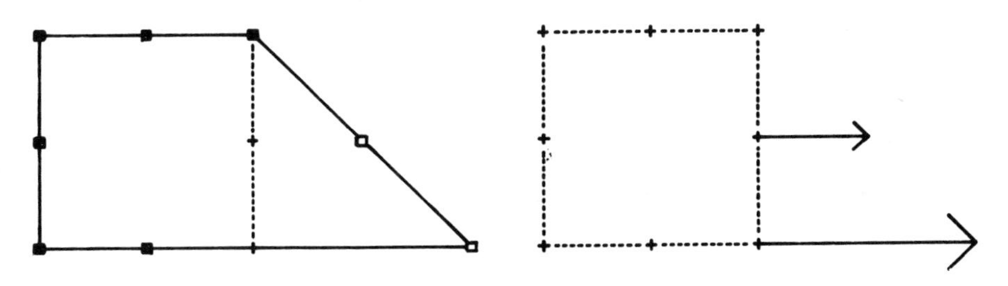

FIGURE 3. Robust fit by repeated medians. Shapes superimposed (left); residuals superimposed on shape 1 (right).

Figures 2 and 3 show that the least squares and robust fits can be very different. The least squares comparison indicates a complicated relationship between the two forms, with differences of various sizes and in various directions at all points. In contrast, the robust fit shows clearly the localized systematic nature of the shape differences. All but two of the homologous pairs are matched closely, and the simple relationship between the shapes is apparent.

Figure 4 shows the comparison of a human skull to that of a chimpanzee. These data are from Sneath (1967) and the example is discussed in detail by Siegel and Benson (1979). It is included here as a demonstration of the graphics enhancement capability. This capability allows for outlines and other details that will not affect the fitting process itself (only the homologous points are used for that) but that will be transformed appropriately in order to produce a useful display.

No dashed lines are seen in Figure 4 because the line capability was not used. However, both the line and enhancement capabilities can be used simultaneously if desired.

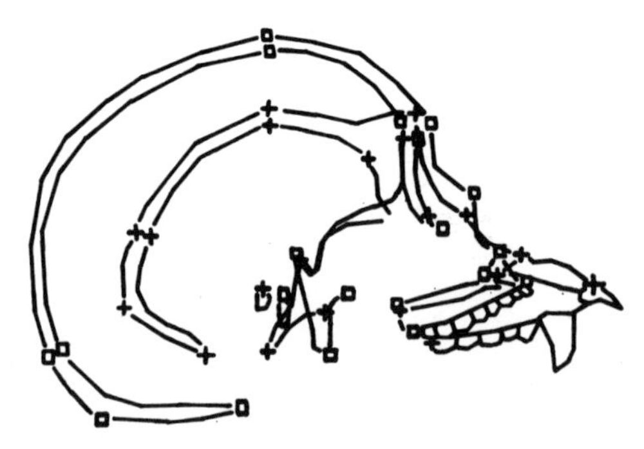

FIGURE 4. Robust comparison of primate skulls, illustrating the enhancement capability.

IV. IMPLEMENTING AND USING THE SYSTEM

To facilitate the use of this program on different computer systems, the program is written in standard FORTRAN but four subroutines must be supplied by the user: INPUT, MOVE, DRAW, and FINISH. These subroutines are used to link to the user's database and graphics capabilities, which vary greatly from one installation to another. Detailed requirements for each subroutine are included in the program listing in the Appendix after each of these four subroutine statements.

The system responds interactively to commands issued by the user. Each command is two letters (additional letters on the same line are ignored) and commands fall into four groups: fit, draw, print, and set. Commands can be issued in any order desired. The fit and set commands can be used to transform shape 2; draw and print commands will be based on the most recent orientation specified (e.g. least squares, robust, or other). For example, the sequence of commands ℓs dr d2 rs pc (one per line) will draw the residuals from the least squares fit, draw shape 2 (still in least squares configuration), and finally print the coordinates of the robust fit.

V. COMMAND DESCRIPTIONS

There are three fit commands: ℓs(least squares), rs (robust), and da(original data). Commands ℓs and rs respectively produce the least squares and robust fits discussed in Section 2. The original data configuration can always be brought back by using da. No printed or graphic output results from these commands, but the internal configuration of shape 2 is transformed accordingly.

There are four draw commands: ds(draw shapes), d1(draw shape 1), d2(draw shape 2), and dr(draw residual vectors). Both shapes are drawn when ds is used; shape 1 alone results when d1 is used; shape 2 alone is drawn when d2 is used. There are three components to a shape drawing: the homologous points (crosses for shape 1 and squares for shape 2), straight lines connecting pairs of points (dashed lines for shape 1, solid lines for shape 2), and enhancements (points and lines that do not affect the fitting process but allow the inclusion of outlines and details to be transformed appropriately). The residual vectors are arrows pointing from the points of shape 1 to the homologous points of shape 2. These vectors are drawn when command dr is given.

Four print commands are included: pc(print coordinates), pp(print parameters), pr(print residuals), and help(print command summary). The command pc prints the coordinates of the points of shape 1, the current values of the transformed homologous points of shape 2, and the current parameter values. The current parameter values a,b,c,d,ρ, and θ, as defined in Section 2, are printed when pp is used. The command pr prints the residual vectors (u(i)-x(i),v(i)-y(i)) together with their lengths, their average length, their root mean square length, and the current parameter values. A summary of valid commands is printed when help is used.

There are four set commands: sp(set parameters), sm (modify parameters), ss(set scale), and se(set point and dash size). These commands prompt you for values to be typed in one per line with five or fewer characters per value, including the required decimal point. You may choose the transformation of shape 2 directly by specifying the parameter values a,b,ρ, and θ using command sp. To modify the current parameter values a,b,ρ, and θ, the command sm will prompt you for changes a´,b´,ρ´, and θ´. The new parameter values will then be a+a´, b+b´, ρρ´, and θ+θ´, unless ρ´=0 in which case ρ will be left unchanged. The overall graphics size can be specified using the ss command; the default size of both shapes will be multiplied by the number entered. The size of the homologous points of both shapes and the spacing between dashes of the lines of shape 1 will both be changed by a factor entered using the se command.

The command en stops execution of the program.

REFERENCES

Huber, P.J. (1980). "Comparison of Point Configurations." Technical Report PJH-1, Department of Statistics, Harvard University.

Siegel, A.F. (1982). "Robust Regression Using Repeated Medians." *Biometrika*, to appear.

Siegel, A.F., and Benson, R.H. (1982). "A Robust Comparison of Biological Shapes." *Biometrics*, to appear.

Sneath, P.H.A. (1967). "Trend-Surface Analysis of Distortion Grids." *Journal of Zoology, Proceedings of the Zoological Society of London 151*, 65-122

Thompson, D'A.W. (1917). "On Growth and Form." First edition, Cambridge University Press.

APPENDIX: FORTRAN PROGRAM

```
c   interactive system for fitting and plotting shapes.
c
c              programmed 10/80 by Andrew F. Siegel.
c
c   commands are listed by entering "help".
c
c
c   variables:
c     n = number of h-points (homologous points) of each shape.
c     m = number of pairs of h-points to be connected.
c         set m=0 if no such lines are desired.
c     nn = 500 = dimension of z, w, and ww (used for
c                enhancement of drawings).
c     ifit = 1, 2, 3, or 4 according to the current fit:
c            original data, least squares, robust, or special
c            choice.
c     coms(16) = table of commands.
c     sf = scale factor. Initially set to one, can be reset to
c          change size of drawings.
c     e = point size and dash spacing.
c     a,b = translation parameters to fit shape 2 to shape 1.
c     rho,theta = magnification and rotation parameters
c                 to fit shape 2 to shape 1.
c     c,d = parameters of linear form of magnification and
c           rotation to fit shape 2 to shape 1.
c           (c=rho*cos(theta),d=rho*sin(theta)).
c     x(n),y(n) = coordinates of h-points of shape 1.
c     uu(n),vv(n) = coordinates of original h-points of
c           shape 2.
c     u(n),v(n) = coordinates of transformed h-points of
c           shape 2.
c     lines(m,2) = pairs of h-points to be connected in
c                  drawing: for example, if lines(k,1)=i and
c                  lines(k,2)=j then the kth line drawn will
c                  connect the ith to the jth h-points of each
c                  shape, using a solid line for shape 1 and a
c                  dotted line for shape 2.
c     z(nn) = piecewise linear enhancements for shape 1.  for
c             example, to draw an outline connecting (a1,b1) to
c             (a2,b2) to...to (ak,bk), store k,a1,b1,...,ak,bk
c             in the vector z().  Additional outlines can be
c             added one after another in this same form, ending
c             finally with a zero.
c     ww(nn) = original piecewise linear enhancements for
c              shape 2.
c     w(nn)  = transformed piecewise linear enhancements for
c              shape 2.
c     ut(n),vt(n) = temporary storage for repeated median
c                   fitting.
c
```

```
c
      dimension x(100),y(100),u(100),v(100),uu(100),vv(100),
     *  lines(200,2),z(500),w(500),ww(500),coms(16),ut(100),
     *  vt(100)
      data coms/'ls','rs','da','ds','d1','d2','dr','pc','pp',
     *         'pr','he','se','sp','sm','ss','en'/
      write(6,1)
1     format(' enter ''help'' for a list of commands.')
      nn=500
      z(1)=0.
      ww(1)=0.
      sf=1.
      e=.0075
      call input(n,nn,m,x,y,uu,vv,lines,z,ww)
      ifit=1
      a=0.
      b=0.
      c=1.
      d=0.
      call coord(n,nn,a,b,c,d,u,v,uu,vv,w,ww)
10    read(5,2)com
2     format(a2)
      if(com.eq.coms(1)) call ls(n,nn,a,b,c,d,ifit,x,y,u,v,uu,
     *  vv,w,ww)
      if(com.eq.coms(2)) call rs(n,nn,a,b,c,d,ifit,x,y,u,v,uu,
     *  vv,w,ww,ut,vt)
      if(com.eq.coms(3)) call data(n,nn,a,b,c,d,ifit,u,v,uu,
     *  vv,w,ww)
      if(com.eq.coms(4)) call ds(0,n,nn,m,ifit,sf,e,x,y,u,v,
     *  lines,w,z)
      if(com.eq.coms(5)) call ds(1,n,nn,m,ifit,sf,e,x,y,u,v,
     *  lines,w,z)
      if(com.eq.coms(6)) call ds(2,n,nn,m,ifit,sf,e,x,y,u,v,
     *  lines,w,z)
      if(com.eq.coms(7)) call dr(n,nn,ifit,sf,e,x,y,u,v,z)
      if(com.eq.coms(8)) call pc(n,a,b,c,d,ifit,x,y,u,v)
      if(com.eq.coms(9)) call pp(a,b,c,d)
      if(com.eq.coms(10)) call pr(n,a,b,c,d,ifit,x,y,u,v)
      if(com.eq.coms(11)) call help
      if(com.eq.coms(12)) call se(e)
      if(com.eq.coms(13)) call sp(n,nn,a,b,c,d,ifit,u,v,uu,vv,
     *  w,ww)
      if(com.eq.coms(14)) call sm(n,nn,a,b,c,d,ifit,u,v,uu,vv,
     *  w,ww)
      if(com.eq.coms(15)) call ss(sf)
      if(com.eq.coms(16)) stop
      go to 10
      stop
      end
c
c
      subroutine input(n,nn,m,x,y,uu,vv,lines,z,ww)
c  reads in coordinate and enhancement data.
      dimension x(n),y(n),uu(n),vv(n),lines(m,2),z(nn),ww(nn)
c
c         ***************************************************
c         the user must supply this subroutine to obtain the
c         number of points, n; the initial coordinates x(),
```

```
c         y(),uu(), and vv(); the values of  m  and  lines(,);
c         and the enhancement values  z()  and  ww().
c         *************************************************
c
      return
      end
c
c
      subroutine move(x,y)
c  moves cursor to  (x,y).
c         *************************************************
c         the user must supply this subroutine to move the
c         graphics cursor to  (x,y)  but draw nothing.
c         generally,  x  and  y  will lie between  Ø  and  1,
c         although these limits can be exceeded.
c         *************************************************
      return
      end
c
c
      subroutine draw(x,y)
c  draws a line to (x,y).
c         *************************************************
c         the user must supply this graphics subroutine to
c         draw a line from the previous cursor location to the
c         point (x,y).  the cursor location should then be
c         updated to  (x,y).
c         *************************************************
      return
      end
c
c
      subroutine finish
c  turns off graphics mode.
c         *************************************************
c         the user must supply this subroutine to turn off
c         graphics mode, if needed.
c         *************************************************
      return
      end
c
c
      subroutine ls(n,nn,a,b,c,d,ifit,x,y,u,v,uu,vv,w,ww)
      dimension x(n),y(n),u(n),v(n),uu(n),vv(n),w(nn),ww(nn)
      ifit=2
      an=n
      sx=Ø.
      sy=Ø.
      su=Ø.
      sv=Ø.
      sux=Ø.
      suy=Ø.
      suu=Ø.
      svx=Ø.
      svy=Ø.
      svv=Ø.
      do 1Ø i=1,n
        sx=sx+x(i)
```

```
          sy=sy+y(i)
          su=su+uu(i)
          sv=sv+vv(i)
          sux=sux+uu(i)*x(i)
          suy=suy+uu(i)*y(i)
          suu=suu+uu(i)*uu(i)
          svx=svx+vv(i)*x(i)
          svy=svy+vv(i)*y(i)
10        svv=svv+vv(i)*vv(i)
        ss=suu-su**2/an+svv-sv**2/an
        c=(sux-su*sx/an+svy-sv*sy/an)/ss
        d=(suy-su*sy/an-svx+sv*sx/an)/ss
        a=(sx-c*su+d*sv)/an
        b=(sy-c*sv-d*su)/an
        call coord(n,nn,a,b,c,d,u,v,uu,vv,w,ww)
        return
        end
c
c
        subroutine rs(n,nn,a,b,c,d,ifit,x,y,u,v,uu,vv,w,ww,ut,
      *   vt)
c   robust fit by repeated medians.
        dimension x(n),y(n),u(n),v(n),uu(n),vv(n),w(nn),ww(nn),
      *   ut(n),vt(n)
        call ls(n,nn,a,b,c,d,ifit,x,y,u,v,uu,vv,w,ww)
        ifit=3
c   get magnification rho
        do 20 i=1,n
          j1=0
          do 10 j=1,n
            if(i.eq.j) go to 10
            j1=j1+1
            ut(j1)=sqrt(((x(j)-x(i))**2+(y(j)-y(i))**2)/
      *                 ((u(j)-u(i))**2+(v(j)-v(i))**2))
10          continue
20        vt(i)=fmed(n-1,ut)
        rho=fmed(n,vt)
c   get rotation theta
        do 40 i=1,n
          j1=0
          do 30 j=1,n
            if(i.eq.j) go to 30
            j1=j1+1
            x1=x(j)-x(i)
            y1=y(j)-y(i)
            u1=u(j)-u(i)
            v1=v(j)-v(i)
            ut(j1)=atan2(u1*y1-v1*x1,u1*x1+v1*y1)
30          continue
40        vt(i)=fmed(n-1,ut)
        theta=fmed(n,vt)
c   get translation vector a,b
        ct=rho*cos(theta)
        dt=rho*sin(theta)
        do 50 i=1,n
          ut(i)=x(i)-ct*u(i)+dt*v(i)
50        vt(i)=y(i)-dt*u(i)-ct*v(i)
        at=fmed(n,ut)
```

```
      bt=fmed(n,vt)
      at=at+ct*a-dt*b
      b=bt+dt*a+ct*b
      a=at
      ctt=ct*c-dt*d
      d=ct*d+dt*c
      c=ctt
      call coord(n,nn,a,b,c,d,u,v,uu,vv,w,ww)
      return
      end
c
c
      subroutine data(n,nn,a,b,c,d,ifit,u,v,uu,vv,w,ww)
c  puts original data back for shape 2.
      dimension uu(n),vv(n),u(n),v(n),ww(nn),w(nn)
      ifit=1
      a=0.
      b=0.
      c=1.
      d=0.
      call coord(n,nn,a,b,c,d,u,v,uu,vv,w,ww)
      return
      end
c
c
      subroutine coord(n,nn,a,b,c,d,u,v,uu,vv,w,ww)
c  puts transformed coordinates into u,v,w for shape 2.
      dimension u(n),v(n),uu(n),vv(n),w(nn),ww(nn)
      do 10 i=1,n
        u(i)=a+c*uu(i)-d*vv(i)
10      v(i)=b+d*uu(i)+c*vv(i)
      i=1
20    j=ww(i)
        w(i)=j
        i=i+1
        if(j.eq.0) return
        do 30 k=1,j
          w(i)=a+c*ww(i)-d*ww(i+1)
          w(i+1)=b+d*ww(i)+c*ww(i+1)
30        i=i+2
        go to 20
      end
c
c
      subroutine ds(nshape,n,nn,m,ifit,sf,e,x,y,u,v,lines,w,z)
c  draw shapes (both if nshape=0, shape 1 only if nshape=1,
c  and shape 2 only if nshape=2).
      dimension x(n),y(n),u(n),v(n),lines(m,2),w(nn),z(nn)
      call desc(ifit)
      call scale(n,s,sf,xm,ym,x,y)
      if(nshape.eq.2) go to 40
c  draw shape 1
      do 10 i=1,n
        p=(x(i)-xm)/s+.5
        q=(y(i)-ym)/s+.5
        call move(p,q+e)
        call draw(p,q-e)
        call move(p+e,q)
```

```
10        call draw(p-e,q)
      call enh(nn,xm,ym,s,z)
      if(m.eq.0) go to 40
      do 30 i=1,m
        i1=lines(i,1)
        i2=lines(i,2)
        x1=(x(i1)-xm)/s+.5
        y1=(y(i1)-ym)/s+.5
        x2=(x(i2)-xm)/s+.5
        y2=(y(i2)-ym)/s+.5
        s2=sqrt((x2-x1)**2+(y2-y1)**2)
        if(s2.lt.4.*e) go to 30
        j=(s2-3.*e)/(2.*e)
        xu=(x2-x1)/s2
        yu=(y2-y1)/s2
        exu=e*xu
        eyu=e*yu
        exu2=2.*exu
        eyu2=2.*eyu
        p=(s2-3.*e-2.*e*float(j))/2.
        x1=x1+p*xu+exu2
        y1=y1+p*yu+eyu2
        do 20 k=1,j
          call move(x1,y1)
          call draw(x1+exu,y1+eyu)
          x1=x1+exu2
20        y1=y1+eyu2
30      continue
40    if(nshape.eq.1) go to 70
c  draw shape 2
      do 50 i=1,n
        p=(u(i)-xm)/s+.5
        q=(v(i)-ym)/s+.5
        call move(p+e,q+e)
        call draw(p+e,q-e)
        call draw(p-e,q-e)
        call draw(p-e,q+e)
50      call draw(p+e,q+e)
      call enh(nn,xm,ym,s,w)
      if(m.eq.0) go to 70
      do 60 i=1,m
        i1=lines(i,1)
        i2=lines(i,2)
        x1=(u(i1)-xm)/s+.5
        y1=(v(i1)-ym)/s+.5
        x2=(u(i2)-xm)/s+.5
        y2=(v(i2)-ym)/s+.5
        s2=sqrt((x2-x1)**2+(y2-y1)**2)
        if(s2.lt.4.*e) go to 60
        xu=(x2-x1)/s2
        yu=(y2-y1)/s2
        call move(x1+2.*e*xu,y1+2.*e*yu)
        call draw(x2-2.*e*xu,y2-2.*e*yu)
60      continue
70    call finish
      return
      end
c
```

```
c
      subroutine enh(nn,xm,ym,s,z)
c  draws enhancements
      dimension z(nn)
      i=1
10    j=z(i)
        i=i+1
        if(j.eq.0) return
        x1=(z(i)-xm)/s+.5
        y1=(z(i+1)-ym)/s+.5
        call move(x1,y1)
        call draw(x1,y1)
        i=i+2
        if(j.eq.1) go to 10
        do 20 k=2,j
          call draw((z(i)-xm)/s+.5,(z(i+1)-ym)/s+.5)
20        i=i+2
        go to 10
      end
c
c
      subroutine dr(n,nn,ifit,sf,e,x,y,u,v,z)
c  draws residual vectors
      dimension x(n),y(n),u(n),v(n),z(nn)
      call desc(ifit)
      call scale(n,s,sf,xm,ym,x,y)
      do 10 i=1,n
        x1=(x(i)-xm)/s+.5
        y1=(y(i)-ym)/s+.5
        u1=(u(i)-xm)/s+.5
        v1=(v(i)-ym)/s+.5
        call move(x1,y1)
        call draw(u1,v1)
        dx=(u1-x1)/8.
        dy=(v1-y1)/8.
        call draw(x1+7.*dx+dy,y1+7.*dy-dx)
        call move(u1,v1)
10      call draw(x1+7.*dx-dy,y1+7.*dy+dx)
      call finish
      return
      end
c
c
      subroutine pc(n,a,b,c,d,ifit,x,y,u,v)
c  prints coordinates and parameters.
      dimension x(n),y(n),u(n),v(n)
      call desc(ifit)
      write(6,1) (x(i),y(i),u(i),v(i),i=1,n)
1     format(' coordinates x,y of shape 1 and u,v of shape 2'/
     *       (1x,4f10.5))
      call pp(a,b,c,d)
      return
      end
c
c
      subroutine pr(n,a,b,c,d,ifit,x,y,u,v)
c  print residuals
      dimension x(n),y(n),u(n),v(n)
```

```
      call desc(ifit)
      write(6,1)
1     format(' residuals: coordinates and lengths.')
      sr=0.
      srr=0.
      do 10 i=1,n
        rx=u(i)-x(i)
        ry=v(i)-y(i)
        r=sqrt(rx**2+ry**2)
        write(6,2) rx,ry,r
2       format(1x,3f10.5)
        sr=sr+r
10      srr=srr+r**2
      sr=sr/float(n)
      srr=sqrt(srr/float(n))
      write(6,3)sr,srr
3     format(/' average residual length = ',f10.5/
     *        ' r.m.s.  residual length = ',f10.5)
      call pp(a,b,c,d)
      return
      end
c
c
      subroutine pp(a,b,c,d)
c  print parameters.
      rho=sqrt(c**2+d**2)
      theta=atan2(d,c)*180./3.141593
      write(6,1)a,b,c,d,rho,theta
1     format(' parameters a, b, c, d, rho, theta (in degrees)'
     *  ,' are'/1x,6f10.5)
      return
      end
c
c
      subroutine help
      write(6,1)
1     format(' valid commands are:'//
     *       5x,'fit:'/10x,'least squares',17x,'ls'/
     *       10x,'robust (repeated median)',6x,'rs'/
     *       10x,'original data',17x,'da'//
     *       5x,'draw:'/10x,'shapes',24x,'ds'/
     *       10x,'shape  1  only',16x,'d1'/
     *       10x,'shape  2  only',16x,'d2'/
     *       10x,'residual vectors',14x,'dr'//
     *       5x,'print:'/10x,'coordinates',19x,'pc'/
     *       10x,'parameters',20x,'pp'/
     *       10x,'residuals',21x,'pr'/
     *       10x,'commands',20x,'help'//
     *       5x,'set:'/10x,'parameters (new)',14x,'sp'/
     *       10x,'parameters (modify)',11x,'sm'/
     *       10x,'scale',25x,'ss'/
     *       10x,'point size, dash spacing',6x,'se'//
     *       5x,'end',32x,'en'/)
      return
      end
c
c
      subroutine se(e)
```

```
c  set point size and dash spacing for drawing shapes.
      write(6,1)
1     format(' please enter desired dash and point size for',
     *  ' drawing.')
      read(5,2)e
2     format(f5.2)
      e=.0075*e
      write(6,3)
3     format(' thank you.')
      return
      end
c
c
      subroutine ss(sf)
c  sets the scale factor to any magnification for drawing.
      write(6,1)
1     format(' please enter desired magnification factor'
     *      ,' for drawing.')
      read(5,2)sf
2     format(f5.2)
      write(6,3)
3     format(' thank you.')
      return
      end
c
c
      subroutine sp(n,nn,a,b,c,d,ifit,u,v,uu,vv,w,ww)
c  sets transformation parameters to any desired values.
      dimension u(n),v(n),uu(n),vv(n),w(nn),ww(nn)
      ifit=4
      write(6,1)
1     format(' desired transformation parameters...'/
     *' enter  a, b, rho, and  theta (degrees), one per line')
      read(5,2)a,b,rho,theta
2     format(f5.2)
      if(rho.eq.0.) rho=1.
      theta=theta*3.141593/180.
      c=rho*cos(theta)
      d=rho*sin(theta)
      call coord(n,nn,a,b,c,d,u,v,uu,vv,w,ww)
      write(6,3)
3     format(' thank you.')
      return
      end
c
c
      subroutine sm(n,nn,a,b,c,d,ifit,u,v,uu,vv,w,ww)
c  modify current values of transformation parameters.
      dimension u(n),v(n),uu(n),vv(n),w(nn),ww(nn)
      ifit=4
      write(6,1)
1     format(' modification of current transformation',
     *  ' parameters...'/' enter  a, b, rho, and  theta',
     *  ' (degrees), one per line.')
      read(5,2)at,bt,rho,theta
2     format(f5.2)
      if(rho.eq.0.) rho=1.
      theta=theta*3.141593/180.
```

```
      ct=rho*cos(theta)
      dt=rho*sin(theta)
      att=at+a*ct-b*dt
      b=bt+a*dt+b*ct
      a=att
      ctt=c*ct-d*dt
      d=c*dt+d*ct
      c=ctt
      call coord(n,nn,a,b,c,d,u,v,uu,vv,w,ww)
      write(6,3)
3     format(' thank you.')
      return
      end
c
c
      subroutine desc(ifit)
c  prints the current type of fit.
      go to (10,20,30,40),ifit
10    write(6,1)
1     format(' original data')
      return
20    write(6,2)
2     format(' least squares fit')
      return
30    write(6,3)
3     format(' robust fit by repeated medians')
      return
40    write(6,4)
4     format(' special fit')
      return
      end
c
c
      subroutine scale(n,s,sf,xm,ym,x,y)
c  finds center and scale for drawing so that shape 1 fits in
c  square from (.25,.25) to (.75,.75), then adjusts by the
c  value of  sf.
      dimension x(n),y(n)
      xmin=x(1)
      xmax=x(1)
      ymin=y(1)
      ymax=y(1)
      do 10 i=1,n
        xmin=amin1(xmin,x(i))
        xmax=amax1(xmax,x(i))
        ymin=amin1(ymin,y(i))
10      ymax=amax1(ymax,y(i))
      s=2.*amax1(xmax-xmin,ymax-ymin)/sf
      xm=(xmin+xmax)/2.
      ym=(ymin+ymax)/2.
      return
      end
c
c
      function fmed(n,a)
c  finds the median of a(1),...,a(n).
c  a more efficient algorithm could be used
c  if  n  is large to decrease execution time.
```

```
      dimension a(n)
      m=n/2+1
      do 20 i=1,m
        k=i
        i1=i+1
        do 10 j=i1,n
10        if(a(j).lt.a(k)) k=j
        at=a(i)
        a(i)=a(k)
20      a(k)=at
      if(n.eq.2*(n/2)) go to 30
      fmed=a(m)
      return
30    fmed=(a(m)+a(m-1))/2.
      return
      end
```

PROJECTION PURSUIT METHODS FOR DATA ANALYSIS*

Jerome H. Friedman
Stanford Linear Accelerator Center
Stanford, California

Werner Stuetzle
Department of Statistics
Stanford University
and
Stanford Linear Accelerator Center
Stanford, California

I. INTRODUCTION

Multivariate analysis can be thought of as a methodology for detection, description and validation of structure in p-dimensional ($p > 1$) point clouds. Classical multivariate analysis relies on the assumption that the observations forming the point cloud(s) have a Gaussian distribution. All information about structure is then contained in the means and covariance matrices, and the well-known apparatus for estimation and inference in parametric families can be brought to bear. The uncomfortable ingredient in this approach is the Gaussianity assumption. The data may be Gaussian with occasional outliers or even the bulk of the data simply might not conform to a Gaussian distribution. The first case is the subject of robust statistics and is not treated here. We discuss methods that do not involve any distributional assumptions. In this case,

*Work supported by the Department of Energy under contract DE-AC03-76SF00515

ISBN 0-12-438180-4

structure cannot be perceived by looking at a set of estimated parameters. An obvious remedy is to look at the data themselves, at the p-dimensional point cloud(s), and to base the description of structure on those views. As perception in more than three dimensions is difficult, the dimensionality of the data first has to be reduced, most simply by projection. Projection of the data generally implies loss of information. As a consequence, multivariate structure does not usually show up in all projections, and no single projection might contain all the information. These points are further illustrated in Chapter 2. It is therefore important to judiciously choose the set of projections on which the model of the structure is to be based. This is the goal of projection pursuit procedures. A paradigm for multivariate analysis based on these ideas is presented in Chapter 3.

By design, projection pursuit methods are ideally suited for implementation or interactive computer graphics systems. The potential of interaction between user and algorithm was convincingly demonstrated in the PRIM-9 system for detection of hypersurfaces and clustering (see Fisherkeller et al [1974]); this system is discussed in Chapter 4. Procedures for multiple regression and multivariate density estimation based on projection pursuit are outlined in Chapters 5 and 6. Common properties of all projection pursuit procedures are discussed in Chapter 7.

2. DETECTION AND DESCRIPTION OF STRUCTURE WITH PROJECTIONS

Our goal is to detect and describe multivariate structure using projections of the data. However, structure, if present,

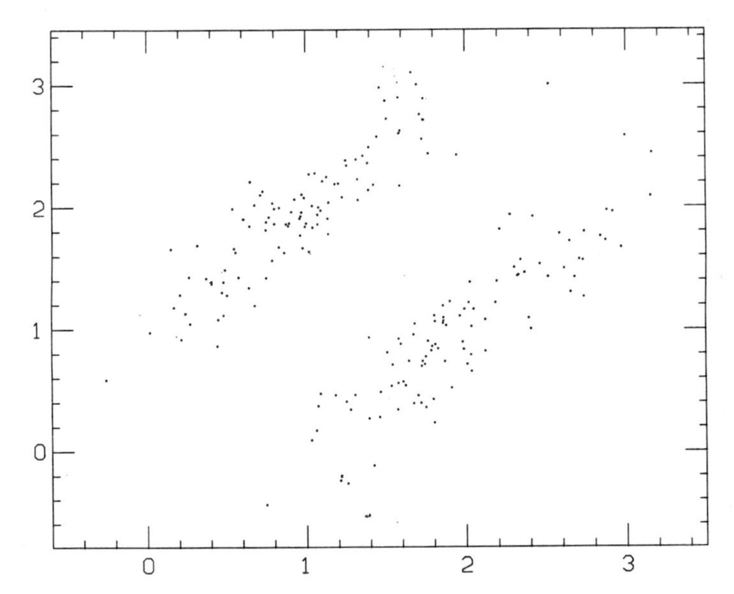

Fig. 1 Structured point cloud in two dimensions

may not be apparent in all projections. This is illustrated by the following examples. Figure 1 shows a point sample drawn from a bivariate distribution. The apparent structure of the point cloud (separation into two clusters) would be revealed by projection onto the subspace spanned by the vector (1, -1), whereas no structure would be apparent in a projection on the subspace spanned by the vector (1, 1).

The data for Figure 2 are generated from the regression model $Y = X_1 + X_2 + \epsilon$ with (X_1, X_2) uniformly distributed in $[-1,1] \times [-1,1]$ and $\epsilon \sim N(0,0.01)$. Figure 2a shows a projection on the two-dimensional subspace spanned by Y and the linear combination $Z = X_1 + X_2$. This projection clearly shows the association between the predictors X_1 and X_2 and the response Y. A similar plot with $Z = X_1 - X_2$, Figure 2b, is clearly less structured.

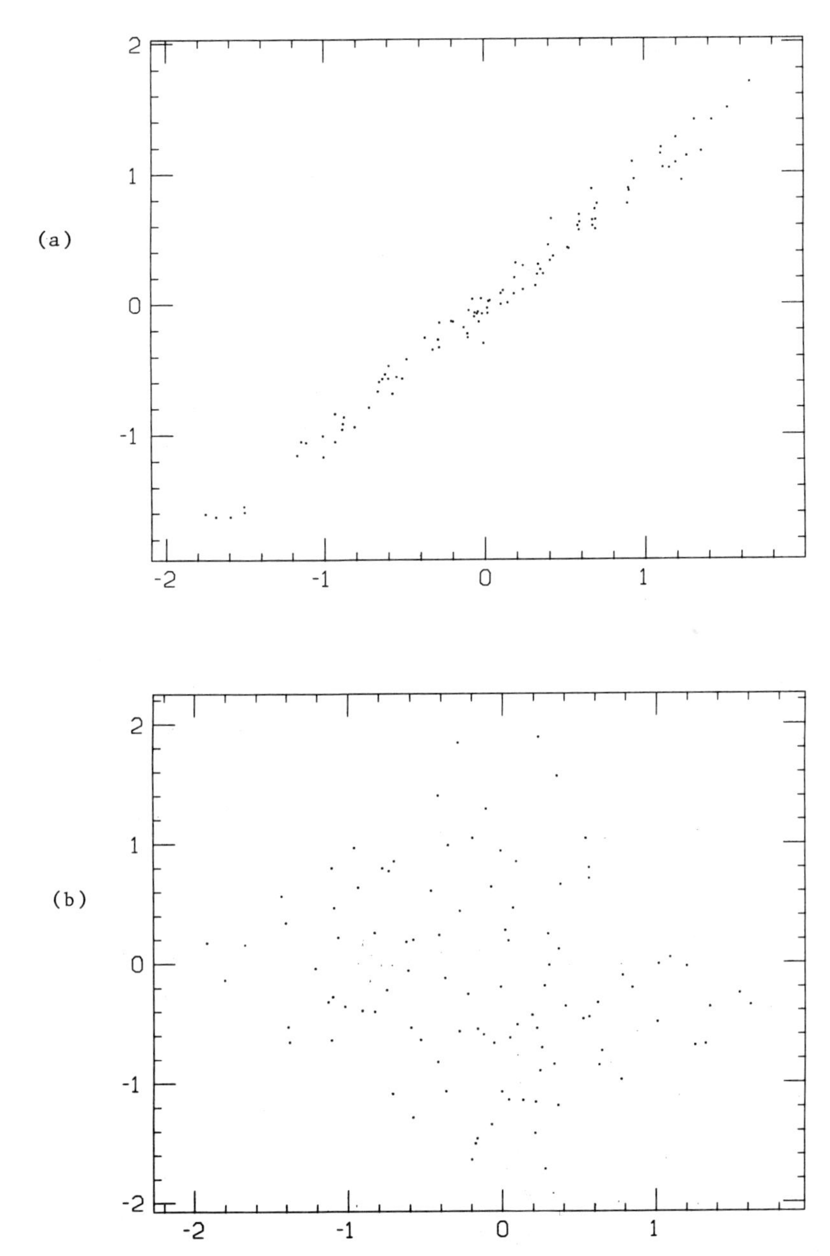

Fig. 2. (a) Projection of data from model $Y=X_1+X_2+\epsilon$ on plane spanned by Y and $Z=X_1+X_2$. (Y is plotted on the vertical axis). (b) Projection of data from model $Y=X_1+X_2+\epsilon$ on plane spanned by Y and $Z=X_1-X_2$.

These examples show that it is important to search for structured projections. This process is called projection pursuit.

It is easy to envision situations where not all the information about the structure is contained in a single projection. Consider the regression example above but with $Y = X_1 \cdot X_2 + \epsilon$. Figures 3a and 3b show two projections with $Z_a = X_1 - X_2$ and $Z_b = X_1 + X_2$. To understand the pictures, note that the simple coordinate transformation $Z_a = X_1 + X_2$, $Z_b = X_1 - X_2$ allows one to express the response as $Y = .25\ (Z_a^2 - Z_b^2)$. It is also interesting to notice that the quadratic dependence on Z_a is washed out due to variability caused by the dependence on Z_b, and vice versa. This suggests that once a structured projection has been found, the structure should be removed so that one obtains a clearer view of what has not yet been uncovered.

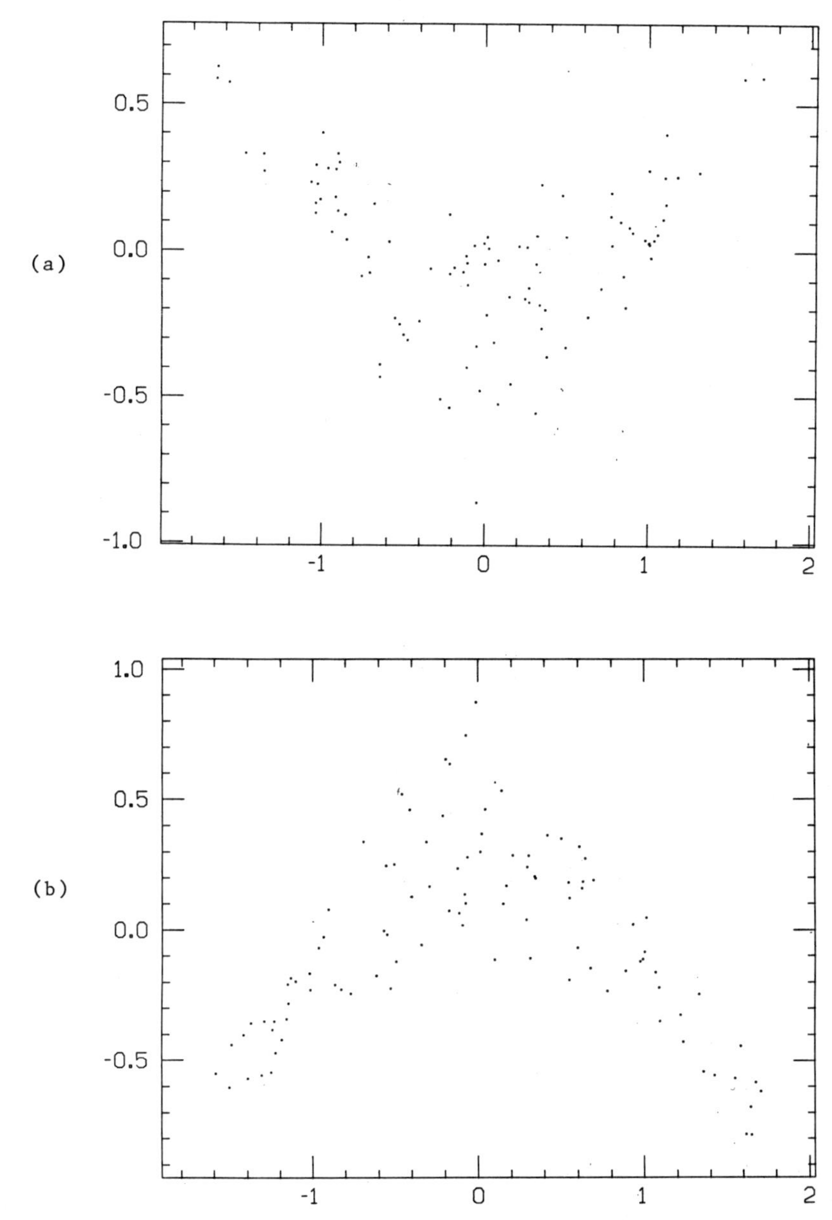

Fig. 3. (a) Projection of data from model $Y=X_1 \cdot X_2+\epsilon$ on plane spanned by Y and $Z_a=X_1+X_2$. (b) Projection of data from model $Y=X_1 \cdot X_2+\epsilon$ on plane spanned by Y and $Z_b=X_1-X_2$.

3. A PROJECTION PURSUIT PARADIGM

The discussion in the previous section motivates the following schema for a class of procedures modeling structure in multivariate data:

(i) Choose an initial model.

<u>Repeat</u>

(ii) Find a projection that shows deviation of the data from the current model, indicating previously undetected structure (<u>Projection Pursuit</u>).

(iii) Change the model to incorporate the structure found in (ii) (<u>Model Update</u>).

<u>Until</u> the current model agrees with the data in all projections.

Such projection pursuit procedures can be implemented in batch mode. In this case, a figure of merit must be defined, which measures the amount of deviation between model and data revealed in a projection. This figure of merit usually is optimized by numerical search, although in some simple cases optimization can be done analytically. If the optimum figure of merit is less than a threshold, data and model are said to agree. Batch implementations of projection pursuit regression and density estimation are described in Sections 5 and 6.

By construction, projection pursuit procedures are ideally suited for implementation on interactive computer graphics systems. Interaction between program and user can help in

- search for interesting projections
- specification of model update
- termination
- interpretation of structure.

Although projection pursuit procedures are useful in batch mode, their full power comes to bear in an interactive environment.

4. THE PRIM-9 SYSTEM

PRIM-9 (Fisherkeller et al, [1974]) is a system for visual inspection of up to nine-dimensional data, mainly intended for detecting clusters and hypersurfaces. It was implemented on an interactive computer graphics system which allows the modification of pictures in real time and thus makes it possible to generate movie-like effects. Its basic set of operations consists of

Projection: The observations can be projected on a subspace spanned by any pair of the coordinates; the projection is shown on a CRT screen.

Rotation: A subspace spanned by any two of the coordinates can be rotated. If the projection subspace and the rotation subspace share a common coordinate, the rotational motion causes the user to perceive a spatial picture of the data as projected on the three-dimensional subspace defined by the coordinates involved. When the user terminates rotation in a particular plane, the old coordinates in that plane are replaced by the current (rotated) coordinates. This makes it possible to look at completely arbitrary projections of the data, not necessarily tied to the original coordinates.

Masking: Subregions of the p-dimensional observation space can be specified, and only points inside the subregion are displayed. Under rotation, points will enter and leave the masked region.

Isolation: Points that are masked out (i.e., not visible) can be removed, thus splitting the data into two subsets.

The first two operations, projection and rotation, allow the user to perform what one might call "manual projection pursuit". Isolation, the splitting of the data set into subsets, provides a rudimentary form of structure removal. When clustering is detected, the clusters can be separated and each of them examined individually. This process can be iterated.

Although several have been implemented (Stuetzle & Thoma [1978], Donoho et al [1981]), systems like PRIM-9 have not yet found widespread use. The main reason has been the price of the necessary computing equipment. The processing power needed to compute rotations at a reasonable update rate is quite high (on the order of 60000 multiplications per second for 1000 observations and 10 updates of the picture per second). Another major cost has been the graphics device, which must have a sufficiently high bandwidth (typically a megabaud). The situation, however, is rapidly changing. New 16-bit microprocessors provide a speed close to that required for an interactive use of projection pursuit procedures. The price of graphics systems, especially raster scan devices, is falling dramatically. The graphics system at SLAC used for the implementation of PRIM-9 cost $175,000 in 1967. Today the price of a comparable system is $15,000.

5. PROJECTION PURSUIT REGRESSION (PPR)

The goal of regression analysis is to find and describe the association between a response variable Y and predictor variables $X_1 \ldots X_p$, using a sample $\{(y_i, \underline{x}_i)\}_{i=1}^{n}$. PPR attempts to construct a model for this association (or, in more classical terms, to estimate $E(Y|\underline{x})$) from the information contained in projections of the data on two-dimensional subspaces spanned by Y and a linear combination $Z = \underline{\alpha} \cdot \underline{X}$. The algorithm exactly follows our projection pursuit paradigm:

(i) Choose an initial model, for example $m_0(\underline{X}) = \text{const.}$

<u>Repeat</u>

(ii) Find a projection that shows deviation of the data from the model, i.e., find a direction such that the current residuals, $r_i = y_i - m(\underline{x}_i)$, show a dependence on $Z = \underline{\alpha} \cdot \underline{X}$

(iii) Describe this dependence by a smooth function $s(Z)$. Update the model:

$m(\underline{X}) \leftarrow m(\underline{X}) + s(\underline{\alpha} \cdot \underline{X})$

<u>Until</u> data and model agree in all projections.

The model after M iterations has the form

$$m(\underline{X}) = m_0(\underline{X}) + \sum_{m=1}^{M} s_m(\underline{\alpha}_m \cdot \underline{X}). \tag{1}$$

PPR allows the modeling of smooth but otherwise completely general regression surfaces. So far, a batch version has been implemented. Such an implementation requires the specification of a figure of merit for projections and a method for summarizing a smooth dependence ("smoother"). Smoothing is generally accomplished by local averaging; the value of the smooth s at a particular point z is obtained by averaging the current residuals r_i for those observations with values of z_i close to z. The size of the neighborhoods within which averaging takes place is called the bandwidth of the smoother. A smoother suitable for use with PPR and guidelines for choosing the bandwidth are described and discussed in Friedman and Stuetzle (1981).

A choice for the figure of merit is suggested by figures 2a and 2b. The (inverse) figure of merit is taken to be the residual sum of squares around the smooth of the current residuals versus $\underline{\alpha} \cdot \underline{X}$. It is small in Figure 2a, where the smooth could closely follow the observations, and large in Figure 2b, where the smooth would be roughly constant. This definition of the figure of merit implies that in each iteration the model is updated along the direction for which the update yields the biggest reduction in residual sum of squares.

As with any stepwise procedure, one needs a criterion for stopping the iteration. Stopping too soon can increase the bias of the estimate, while not stopping soon enough can unduly increase its variance. "Optimal" termination of stepwise procedures has been studied (see Stone, [1981]); these methods can be applied here. In practice, the iteration is usually terminated subjectively, based on differences between successive values of the residual sum of squares. In addition, graphical inspection of $s_m(\underline{\alpha}_m \cdot \underline{X})$ can be used to judge whether the corresponding term should be included in the model. If the graph of s_m shows a noisy pattern with no systematic tendency, then its inclusion can only increase the variability of the estimate. On the other hand, a definite dependence indicates that s_m deals with an inadequacy of the present model.

The following example illustrates the operation of PPR. A sample of 200 observations was generated according to the model

$$Y = 10 \sin(\pi X_1 X_2) + 20(X_3-0.5)^2 + 10X_4+5X_5+0X_6+\epsilon$$

with $(X_1,\ldots,X_6)$ uniformly distributed in $[-1,1]^6$ and $\epsilon \sim N(0,1)$. Figure 4a shows Y plotted against the best single predictor,

X_4, and the corresponding smooth. (The response Y is plotted on the vertical axis, X_4 on the horizontal axis. The "+" symbols represent data points, numbers indicated more than 1 data point. The smooth is represented by the "*" symbols.) Figure 4b shows Y plotted against the linear combination $\underline{\alpha}_1 \cdot \underline{X}$ found in the first iteration with $\underline{\alpha}_1$ = (0.41, 0.51, -0.04, 0.69, 0.31, 0.0). The association is seen to be approximately linear. The model after the first iteration thus is a plane which, in this case, closely coincides with the least squares plane through the data. Figure 4c shows the residuals from this model plotted against the second linear combination $\underline{\alpha}_2 \cdot \underline{X}$ found by the algorithm, with $\underline{\alpha}_2$ = (-0.14, 0.0, 0.99, 0.04, 0.0, -0.03). This iteration is seen to incorporate the quadratic dependence of the response on X_3 into the model. Figure 4d shows the residuals after two iterations plotted against the third linear combination with $\underline{\alpha}_3$ = (.0.70, 0.72, 0.01, 0.03, 0.02, 0.00). Figure 4e shows the residuals after three iterations plotted against the fourth linear combination, with $\underline{\alpha}_4$ = (0.80, -0.59, -0.10, 0.04, 0.01, 0.0). The last two iterations are seen to model the interaction term $\sin(\pi X_1 X_2)$. A further iteration failed to substantially improve the model.

For a more complete discussion of PPR and additional examples, see Friedman and Stuetzle [1981].

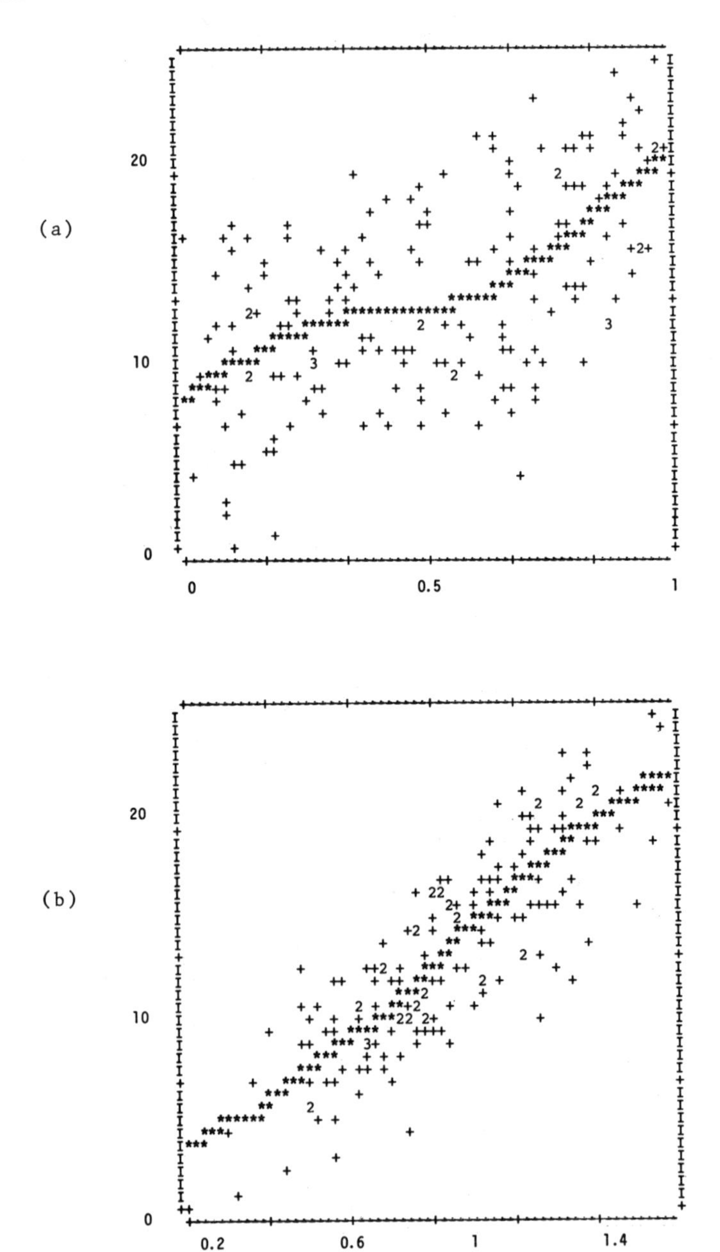

Fig. 4. (a) Scatterplot of Y vs. X_4. (b) Scatterplot of Y vs. $\underline{\alpha}_1 \cdot \underline{X}$. (c) Scatterplot of $Y - s_1(\underline{\alpha}_1 \cdot \underline{X})$ vs. $\underline{\alpha}_2 \cdot \underline{X}$. (d) Scatterplot of $Y - s_1(\underline{\alpha}_1 \cdot \underline{X}) - s_2(\underline{\alpha}_2 \cdot \underline{X})$ vs. $\underline{\alpha}_3 \cdot \underline{X}$. (e) Scatterplot of $Y - s_1(\underline{\alpha}_1 \cdot \underline{X}) - s_2(\underline{\alpha}_2 \cdot \underline{X}) - s_3(\underline{\alpha}_3 \cdot \underline{X})$ vs. $\underline{\alpha}_4 \cdot \underline{X}$.

(c)

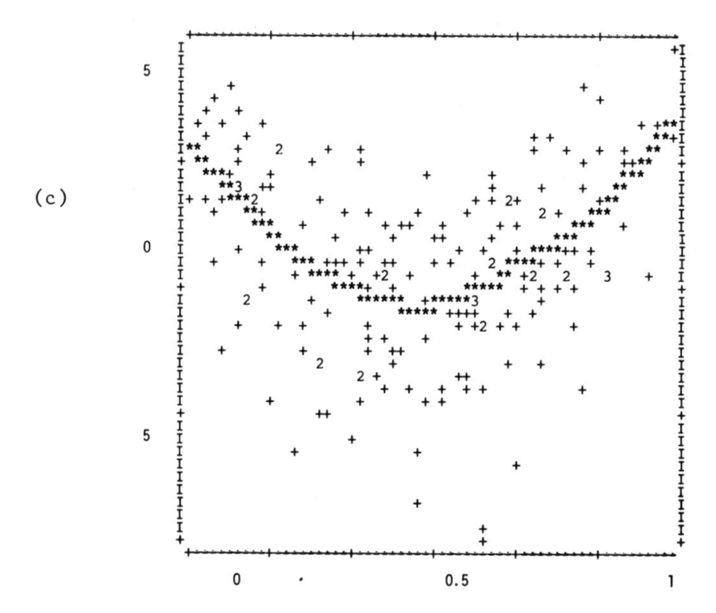

(d)

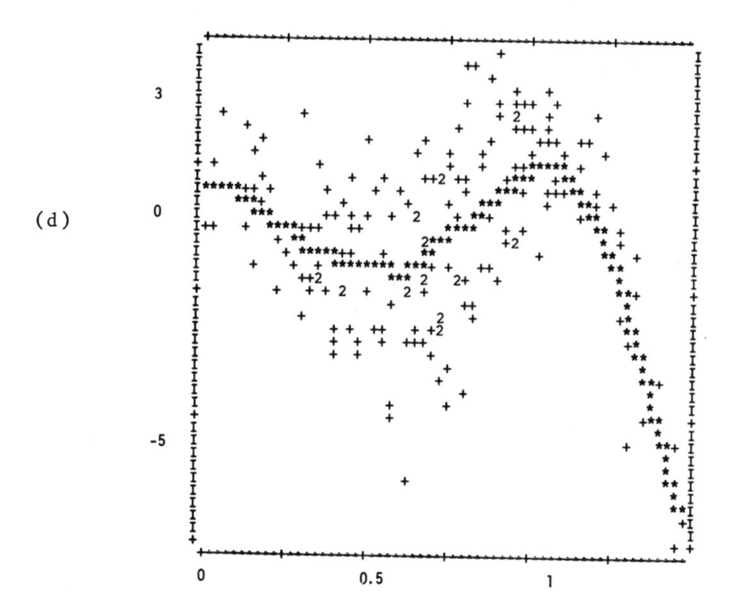

Fig. 4c,d

(e)

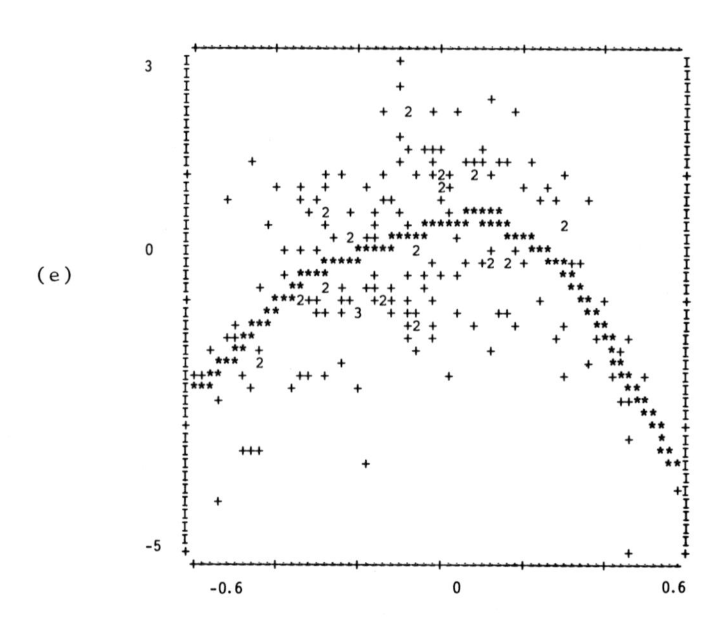

Fig. 4e

6. PROJECTION PURSUIT DENSITY ESTIMATION (PPDE)

The goal of density estimation is to estimate the multivariate distribution of a random vector $\underline{X}$ on the basis of an i.i.d. sample $\underline{x}_1 \ldots \underline{x}_n$. Our procedure again follows the projection pursuit paradigm:

(i) Choose an initial model for the density, for example, m_0 = multivariate normal with sample mean and covariance matrix.

Repeat

(ii) Find a projection that shows deviation of the data from the model; i.e., find a direction such that $m(\underline{\alpha} \cdot \underline{X})$, the model marginal along $\underline{\alpha}$, differs from $p(\underline{\alpha} \cdot \underline{X})$, the (estimated) data marginal along $\underline{\alpha}$.

(iii) Define an "augmenting function" $f(\underline{\alpha} \cdot \underline{X})$ as the quotient of data and model marginals

$$f(\alpha \, . \, X) = \frac{p(\underline{\alpha} \, . \, \underline{X})}{m(\underline{\alpha} \, . \, \underline{X})}$$

Update the model so that it and the data agree in the marginal along $\underline{\alpha}$:

$$m(\underline{X}) \leftarrow m(\underline{X}) \cdot f(\underline{\alpha} \, . \, \underline{X}).$$

Until data and model agree in all projections.

The model after M steps of the iteration is of the form

$$m(\underline{X}) = m_0(\underline{X}) \cdot \prod_{m=1}^{M} f_m(\underline{\alpha}_m \, . \, \underline{X}). \qquad (2)$$

In step (iii) of the algorithm, the marginal of the data along $\underline{\alpha}$ must be estimated and the marginal of the current model must be computed. The data marginal presents no problem. It can be estimated by projecting the data onto $\underline{\alpha}$ and using a one-dimensional kernel or near neighbor estimate. The analytic computation of the model marginal can be very difficult because it requires a (p-1)-dimensional integration. We perform the integration by Monte Carlo, generating a sample from the model and proceeding as in the estimation of the data marginal.

As in the case of PPR, only a batch version of PPDE has so far been implemented. At each iteration, the direction $\underline{\alpha}$ is chosen such that the update of the current model yields the largest improvement in goodness-of-fit as measured by the likelihood of the sample. Termination rules are analogous to those used in PPR.

The following example illustrates the operation of PPDE. The data for the example are the concentration levels of four hormones in blood measurements of 256 children. The purpose of applying PPDE to these data is to determine if a Gaussian distribution represents a reasonable approximation to the data density. Figures 5a-5d compare the experimental data to a Monte Carlo sample drawn from a Gaussian density with the sample mean and covariance, as projected onto each of the measurement coordinates. The histogram of the experimental data is drawn with solid lines; the histogram of the Monte Carlo data is indicated by "*" symbols. Inspection of these projections indicates that although there are possibly some discrepancies, a Gaussian density might be a reasonable approximation to the data.

Figures 6a-6e show results for three iterations of PPDE. The solution direction $\underline{\alpha}_1$ associated with the first iteration is mainly a combination of the second and third coordinate measurements. The data distribution (Figure 6a) is seen to be somewhat skew and more peaked than the corresponding Gaussian. The discrepancy between the data and the Gaussian model is much more pronounced in this projection than on any of the original coordinate measurements. Figure 6b plots the augmenting function $f_1(\underline{\alpha}_1 \cdot \underline{X})$.

The second linear combination $\underline{\alpha}_2$ mainly involves the third and fourth coordinates. The principal difference between the current model $p_1(\underline{X})$ and the data is seen to be a substantial skewness to the left (Figure 6c). Figure 6d shows the corresponding augmenting function f_2. The linear combination associated with the third projection mostly involves the first and second coordinates. Although this iteration is trying to account for an apparent additional skewness of the data (Figure 6e), the effect is seen to be relatively small and perhaps not significant.

(a)

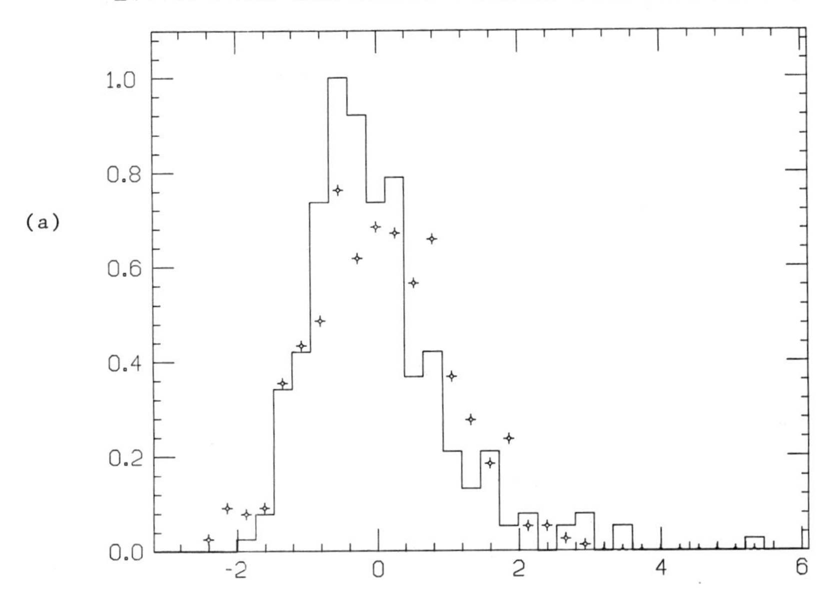

(b)

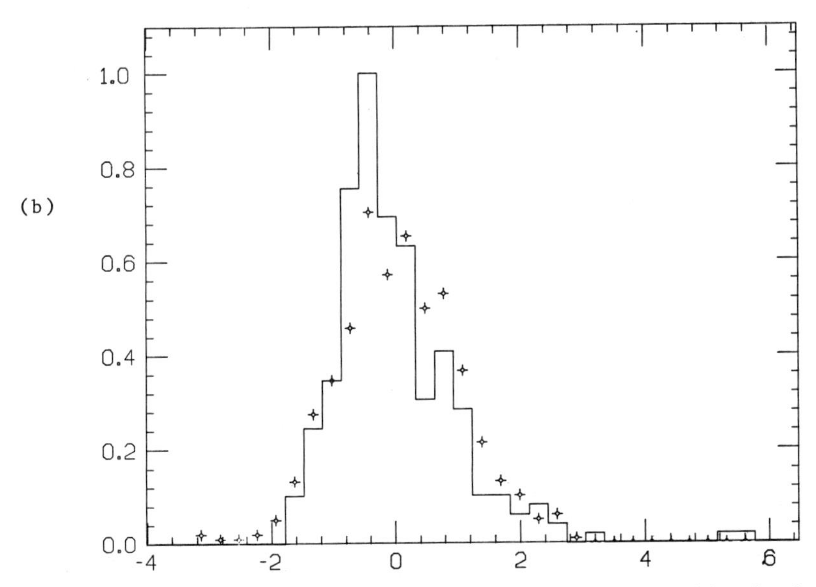

Fig. 5. (a) Hormone data: Histogram of variable 1 with Gaussian Monte Carlo superimposed. (b) Hormone data: Histogram of variable 2 with Gaussian Monte Carlo superimposed. (c) Hormone data: Histogram of variable 3 with Gaussian Monte Carlo superimposed. (d) Hormone data: Histogram of variable 4 with Gaussian Monte Carlo superimposed.

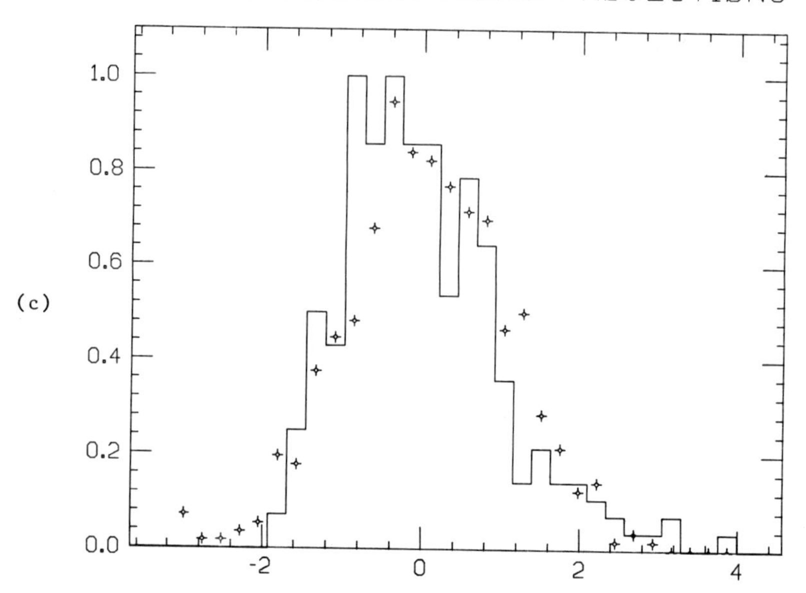
DATA AND CURRENT MØDEL PRØJECTIØNS
(c)
1.0
0.8
0.6
0.4
0.2
0.0
-2
0
2
4

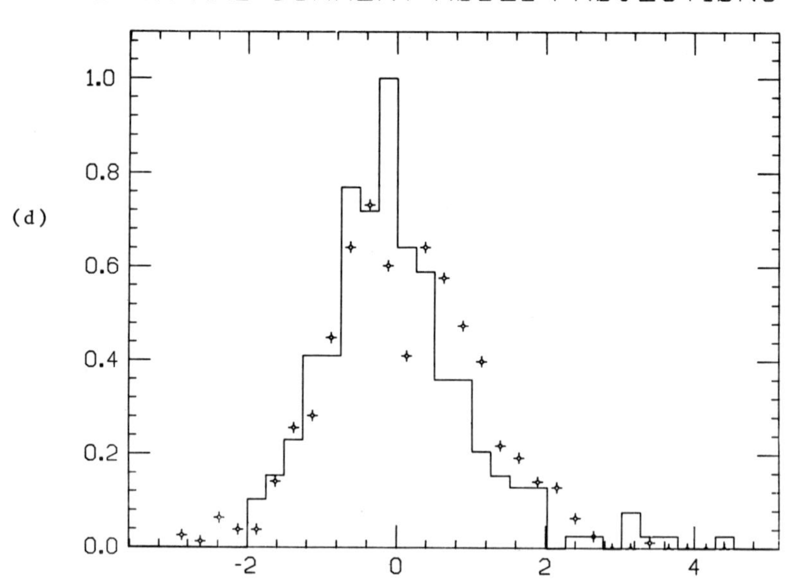
DATA AND CURRENT MØDEL PRØJECTIØNS
(d)
1.0
0.8
0.6
0.4
0.2
0.0
-2
0
2
4

Fig. 5c,d

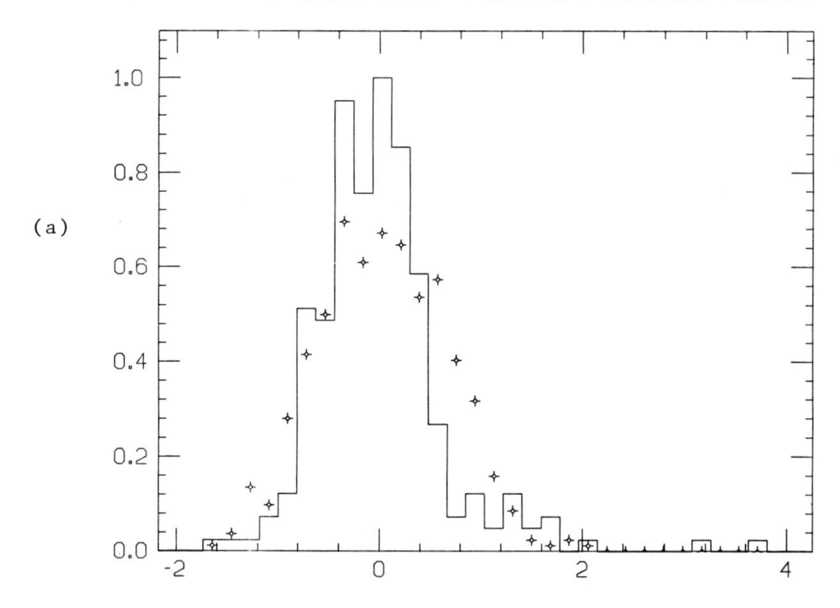

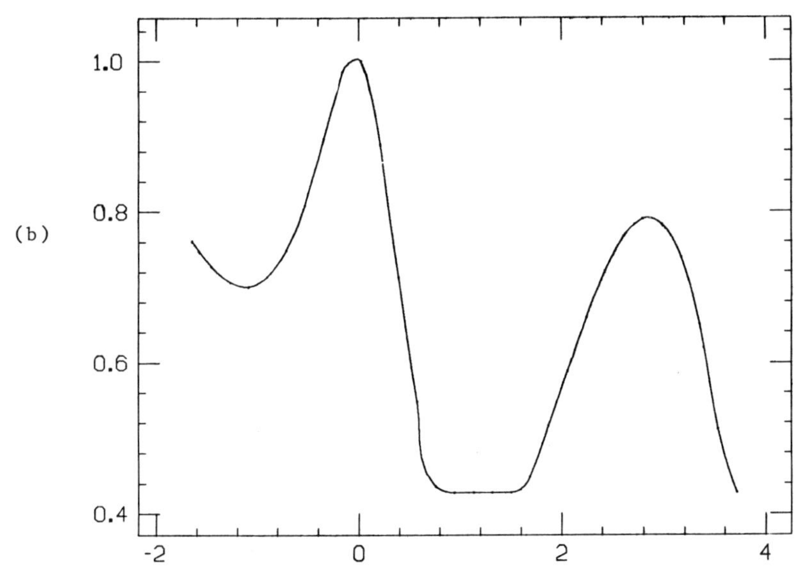

Fig. 6. Hormone Data: (a) Histogram of 1st solution linear combination $\underline{\alpha}_1$=(0.02,0.08,-0.59,0.14) with Gaussian model superimposed. (b) Augmenting function along 1st solution linear combination α_1 (0.02,0.80,-0.59,0.14). (c) Histogram of 2nd so-solution linear combination $\underline{\alpha}_2$=(-0.09,0.18,-0.45,-0.87) with current model Monte Carlo superimposed. (d) Augmenting function along second linear combination α_2=(-0.09,0.18,-0.45,-0.87). (e) Histogram of third solution linear combination α_3=(0.45, 0.88, 0.16,-0.02) with current model Monte Carlo superimposed.

(c)

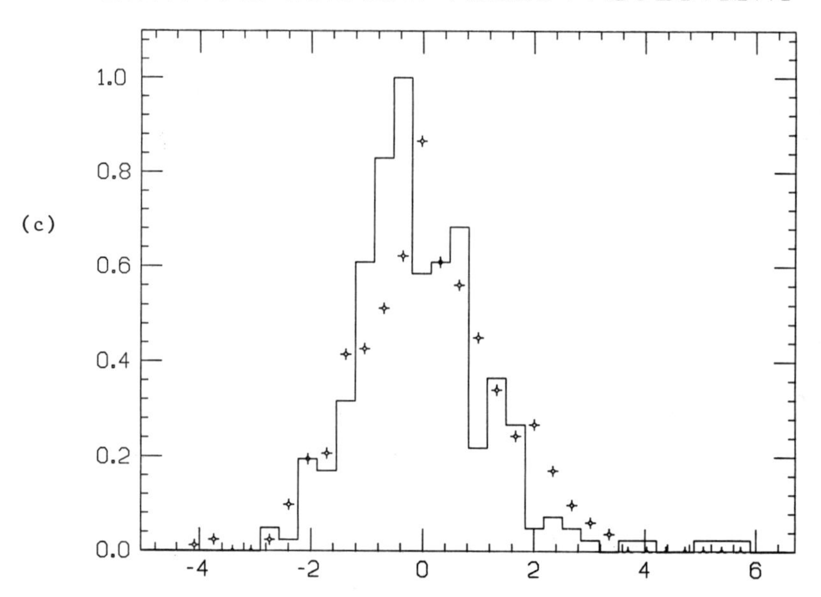
DATA AND CURRENT MØDEL PRØJECTIØNS
1.0
0.8
0.6
0.4
0.2
0.0
-4
-2
0
2
4
6

(d)

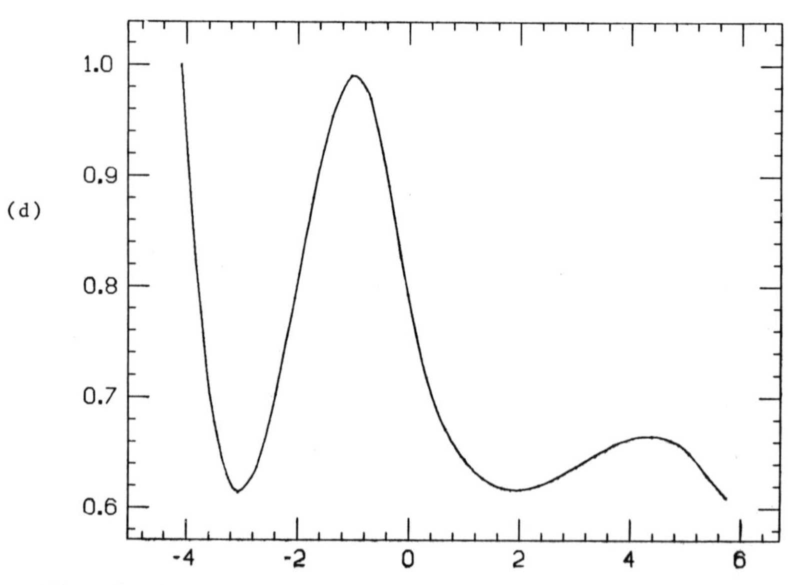
AUGMENTING FUNCTIØN
1.0
0.9
0.8
0.7
0.6
-4
-2
0
2
4
6

Fig. 6c,d.

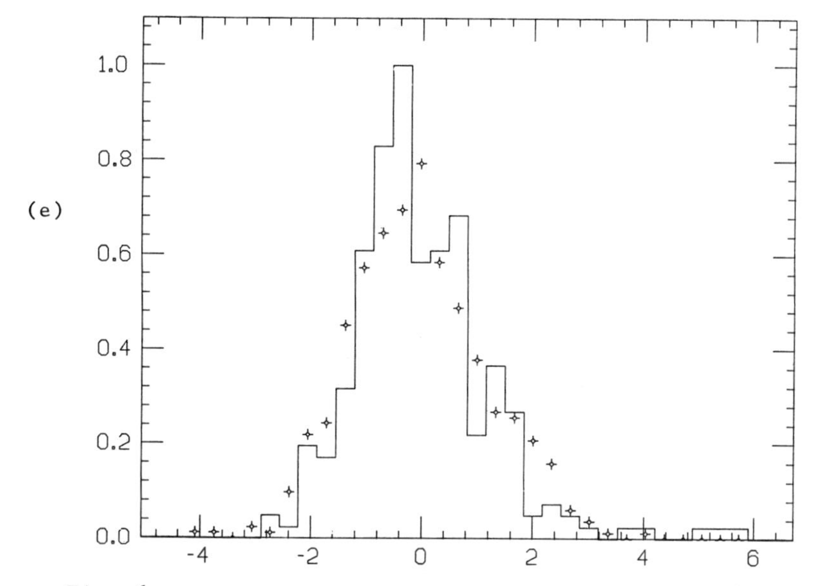

Fig. 6e.

Application of PPDE to these data reveals that a Gaussian model provides a considerably less adequate description than indicated by the coordinate projections alone. The associated graphics gives some insight into the nature of the nonnormality of the data.

7. DISCUSSION

All projection pursuit procedures share some common advantages:

- Since all estimation is carried out in a univariate setting, the large bias of kernel or near neighbor estimates in high dimensions can often be avoided.
- PP procedures do not require specification of a metric in the observation space.
- Bias is encountered with stepwise procedures when many terms are required to provide a good representation of the model

underlying the data, but only a few can be estimated due to insufficient sample size. In these cases, it is important that the first few terms be able to approximate a wide variety of functions so that the most salient features of the data can be modeled. In the limit $M \to \infty$, any regression function can be represented by (1), and any density can be represented by (2) (independent of the initial model), but even for moderate M, functions of those types constitute rich classes. In addition, the choice of the initial model permits the user to introduce any knowledge (s)he may have concerning the data, thereby allowing a further reduction in bias.

- As a data-analytic tool, projection pursuit procedures provide a set of directions $\underline{\alpha}_1 \ldots \underline{\alpha}_M$ for exploring the differences between the initial model and the data. The fact that at each stage the direction is chosen, for which the current model least adequately describes the data, makes them good candidates for that purpose. A graphical comparison of the projections of model and data, along with knowledge of the initial model, can yield considerable insight into the multivariate data distribution. Pictorial representations of each of the augmenting functions s_m, respectively f_m, along with the particular directions over which they vary, can also be quite informative since it is these functions that actually comprise the model.

There are situations in which projection pursuit procedures can be expected not to perform well. Examples of regression functions requiring a large number of terms in equation (1) are those with multiple peaks. Examples of unfavorable density functions are those with highly concave isopleths or with sperically nested isopleths of the same density value.

In addition to regression and density estimation, the projection pursuit paradigm can be applied to the problems of classification and robust estimation of covariance matrices. All projection pursuit procedures use the same set of basic operations, projection pursuit and model update. This should allow the design of an interactive system for analysis of multivariate data that covers a wide range of problems and yet is easy to learn and simple to operate.

1. Fisherkeller, M.A., Friedman, J.H., Tukey, J.W., PRIM-9: An Interactive Multidimensional Data Display and Analysis System, SLAC-PUB-1408, April 1974.
2. Stuetzle, W., Thoma, M., PRIMS-ETH: A Program for Interactive Graphical Data Analysis, Research Report No. 19, Fachgruppe fur Statistik ETH Zurich, October 1978.
3. Donoho, D., Huber, P.J., Thoma, M., (1981). The Use of Kinematic Displays to Represent High Dimensional Data. To appear in Computer Science and Statistics, Proceedings of the 14th Annual Symposium on the Interface.
4. Friedman, J.H., Stuetzle, W. (1981). Projection Pursuit Regression. To appear in JASA.
5. Stone, C.J. (1981). Admissible Selection of an Accurate and Parsimonious Normal Linear Regression Model. Ann. Statist. 9, (to appear).

INFLUENCE FUNCTIONS

AND REGRESSION DIAGNOSTICS

Roy E. Welsch[1]

Sloan School of Management

Massachusetts Institute of Technology

Cambridge, Massachusetts

I. INTRODUCTION

Influential-data diagnostics are becoming an accepted part of data analysis. Recent work is described in Belsley, Kuh, and Welsch (1980), Cook and Weisberg (1980), and Pregibon (1981). In this paper we show how these diagnostic techniques are connected with the ideas of qualitative robustness described by Hampel (1968, 1974) and the concept of bounded-influence regression as developed by Krasker and Welsch (1979).

II. ASYMPTOTIC INFLUENCE FUNCTIONS

We begin with the classical linear regression model

$$y_i = x_i^T\beta + \varepsilon_i \qquad i = 1, \ldots, n$$

where y_i is the ith observation on the response variable, x_i^T is the p-dimensional row vector containing the ith observation

[1]Supported, in part, by D.O.E. grant EX-76-A-01-2295 and NSF grant 7726902-A04-MCS to the M.I.T. Center for Computational Research in Economics and Management Science.

ISBN 0-12-438180-4

on the p explanatory variables, β is the p-vector of unknown parameters, and the disturbances ε_i are i.i.d. and independent of the x's.

The least-squares linear regression estimates, b and s, may be written as the solution to

$$\frac{1}{n}\sum_{i=1}^{n} x_i(y_i - x_i^T b) = 0 \tag{1}$$

$$\frac{1}{n}\sum_{i=1}^{n} (y_i - x_i^T b)^2 = \left(\frac{n - p}{n}\right) s^2 \tag{2}$$

or equivalently

$$\int_{x,y} x(y - x^T b)\, dF_n(x,y) = 0 \tag{3}$$

$$\int_{x,y} (y - x^T b)^2 dF_n(x,y) = \left(\frac{n - p}{n}\right) s^2 \tag{4}$$

where F_n is the empirical distribution function.

Assume now that F_n converges to F, the true underlying common distribution function of the observations. Let $T(F_n)$ denote the functional for b defined on the space of empirical distribution functions by (3) and $S(F_n)$ the functional for s defined by (4). We assume that T(F) equals the limit in probability of $T(F_n)$ with a similar assumption for $S(F_n)$. Then (3) and (4) give us the definition of T(F) and S(F) with F_n replaced by F, except for the (n - p)/n in (4) which we replace by 1.

Hampel (1974) showed that the suitably normed limiting influence of adding one more observation to a very large sample could be expressed as

$$\Omega(x,y,F,T) = \lim_{\varepsilon \to 0} \frac{T[(1-\varepsilon)F + \varepsilon\delta_{x,y}] - T(F)}{\varepsilon} \tag{5}$$

where $\delta_{x,y}$ denotes the point mass 1 at (x,y). This is often called the influence curve or function. If T is sufficiently regular,

$$\sqrt{n}\,(T(F_n) - T(F)) = \frac{1}{\sqrt{n}} \sum_{i=1}^{n} \Omega(x_i,y_i,F,T) + \ldots \tag{6}$$

and in many cases $\sqrt{n}\,(T(F_n) - T(F))$ will be asymptotically normal with mean zero and asymptotic variance matrix

$$V = \int \Omega(x,y,F,T)\Omega^T(x,y,F,T)\,dF(x,y). \tag{7}$$

Hampel (1974) also defined the gross-error sensitivity as

$$\gamma_1 = \sup_{x,y} (\Omega^T(x,y,F,T)\Omega(x,y,F,T))^{\frac{1}{2}} \tag{8}$$

in order to provide a measure of the maximum possible influence of any observation (x,y) on the coefficients.

Other definitions of sensitivity are possible. For example,

$$\gamma_2 = \sup_{x,y} |x^T\Omega(x,y,F,T)| \tag{9}$$

which measures the influence of (x,y) on the fitted value. More generally we might be interested in the influence on a linear combination of the parameters (e.g. a predicted value)

$$\sup_{x,y} |\ell^T\Omega(x,y,F,T)|.$$

The fit influence (9) is just a special case. Note that if we want to consider all linear combinations

$$\sup_{x,y} \sup_{\ell} \frac{|\ell^T \Omega(x,y,F,T)|}{(\ell^T \ell)^{\frac{1}{2}}},$$

we get γ_1.

Since there is stochastic variability in our estimates, we may not be overly interested in the influence of (x,y) as measured by $\ell^T \Omega(x,y,F,T)$ unless it were large relative to the standard error, $(\ell^T V \ell)^{\frac{1}{2}}$, of the linear combination. Since

$$\sup_{\ell} \frac{|\ell^T \Omega(x,y,F,T)|}{(\ell^T V \ell)^{\frac{1}{2}}}$$

$$= [\Omega^T(x,y,F,T) V^{-1} \Omega(x,y,F,T)]^{\frac{1}{2}} \equiv J(x,y,F,T), \tag{10}$$

we define the sensitivity as

$$\gamma_3 = \sup_{x,y} J(x,y,F,T).$$

It is easy to show from (7) that

$$\gamma_3 \geq E_{x,y}[J(x,y,F,T)]$$

$$= \text{trace } E_{x,y}(V^{-1}\Omega(\cdot)\Omega^T(\cdot))^{\frac{1}{2}} = p^{\frac{1}{2}}. \tag{11}$$

This measure of sensitivity forms the basis for the bounded-influence regression estimators developed by Krasker and Welsch (1979).

III. INFLUENCE DIAGNOSTICS

The influence function measures what happens when we add one observation (x,y) to a very large sample. It would be useful to be able to compute the influence function because we would then gain some understanding of how our estimates are influenced by single observations. In practice, of course, we do not have a very large sample. Is there an approximation to $\Omega(x,y,F,T)$ that will give us similar information? There are several ways to approximate $\Omega(x,y,F,T)$, and we will discuss two of the promising ones here.

Since we are often interested in how the ith observation (x_i, y_i) influences our least-squares estimate, it seems natural to consider adding (x_i, y_i) to a sample of size n-1 (the remaining data). If we put F_{n-1} into (3) and evaluate the limit in (5) we obtain

$$\Omega(x_i,y_i,F_{n-1},b(i))$$

$$= (n - 1)(X^T(i)X(i))^{-1}x_i(y_i - x_i^Tb(i)) \tag{12}$$

where (i) denotes deletion of the ith row of data and we have replaced $T(F_{n-1})$ by $b(i)$ in order to denote which observations form the set of size n-1. The quantity $y_i - x_i^Tb(i)$ is called the predicted residual and

$$y_i - x_i^Tb(i) = e_i/(1 - h_i) \tag{13}$$

where $h_i = x_i^T(X^TX)^{-1}x_i$ and $e_i = y_i - x_i^Tb$. Thus

$$\Omega(x_i,y_i,F_{n-1},b(i)) = (n - 1)(X^T(i)X(i))^{-1}x_ie_i/(1 - h_i)$$

$$= (n - 1)(X^TX)^{-1}x_ie_i/(1 - h_i)^2. \tag{14}$$

The definition of sensitivity based on (10) provides a good way to scale the influence function and invariance to nonsingular transformations of the X-matrix. To use it we require an approximation to (7). Now since

$$V = E_xE_{y|x}(\Omega(\cdot)\Omega^T(\cdot))$$

we use the empirical distribution function F_{n-1} with the ith observation omitted to obtain

$$V(i) = (n - 1)\cdot$$

$$\sum_{k \neq i} (X^T(i)X(i))^{-1} x_k x_k^T (X^T(i)X(i))^{-1} E_{y|x_k}(y - x_k^T b(i))^2. \tag{15}$$

A reasonable estimate for $E_{y|x_k}(y - x_k^T b(i))^2$ would be to use (4) with F_{n-1} giving

$$s^2(i) = \frac{1}{(n - 1) - p} \sum_{k \neq i} (y_k - x_k^T b(i))^2 \tag{16}$$

which implies that

$$V(i) = (n - 1)s^2(i)(X^T(i)X(i))^{-1}. \tag{17}$$

Combining (14) and (17) gives a finite sample approximation to (10) of

$$(n - 1)^{\frac{1}{2}}(x_i^T(X^T(i)X(i))^{-1}x_i)^{\frac{1}{2}} \frac{|e_i|}{s(i)(1 - h_i)}$$

$$= (n - 1)^{\frac{1}{2}}\left(\frac{h_i}{1 - h_i}\right)^{\frac{1}{2}} \frac{|e_i|}{s(i)(1 - h_i)} \tag{18}$$

$$= (n - 1)^{\frac{1}{2}}\left(\frac{h_i}{1 - h_i}\right)^{\frac{1}{2}} \frac{|y_i - x_i^T b(i)|}{s(i)}. \tag{19}$$

When a constant column is present in X, the Malhalanobis distance (Rao, 1973) between x_i^T and $\bar{x}^T(i)$ is

$$\frac{(n - 2)}{(n - 1)} \left[\frac{n(h_i - \frac{1}{n})}{1 - h_i}\right]^{\frac{1}{2}}$$

which means that (19) is essentially the Malhalanobis distance multiplied by the predicted residual and divided by an estimate of σ.

A different interpretation of $h_i/(1 - h_i)$ is given by the formula (Huber, 1981)

$$\hat{y}_i = (1 - h_i)x_i^T b(i) + h_i y_i$$

which shows that $h_i/(1 - h_i)$ is the fraction of the fitted value, $\hat{y}_i$, due to y_i divided by the fraction due to the predicted value, $x_i b(i)$.

Another approximation to the influence function (Tukey, 1970) can be achieved by omitting the limit, and letting $\varepsilon = 1/n$ and $F = F_{n-1}$ in (5), giving

$$n[T((1 - \frac{1}{n})F_{n-1} + \frac{1}{n}\delta_{x_i,y_i}) - T(F_{n-1})].$$

$$= n[T(F_n) - T(F_{n-1})].$$

This may be viewed as a type of secant approximation to the limit in (5). In the least-squares case we obtain

$$n(b - b(i)) = \frac{n(X^TX)^{-1}x_i e_i}{(1 - h_i)} \tag{20}$$

$$= n(X^TX)^{-1}x_i(y_i - x_i^T b(i)).$$

Often (20) is preferred to the finite sample approximation, (14), because it is easier to interpret.

Using the asymptotic formula (7) with (20) and F_{n-1} to approximate the covariance, and (16) to approximate σ^2, we have

$$\tilde{V}(i) = \frac{n^2}{n - 1} s^2(i)(X^TX)^{-1}(X^T(i)X(i))(X^TX)^{-1}. \tag{21}$$

If we put (20) and (21) into (10), then we get (18) again. When (17) is used to approximate (7), and (17) and (20) are substituted into (10), we obtain

$$(n-1)^{\frac{1}{2}}\left(\frac{h_i}{1-h_i}\right)^{\frac{1}{2}} \frac{|e_i|}{s(i)} \;. \tag{22}$$

While (18) seems to be very natural measure of influence, Malhalanobis distance and predicted residuals are not well-known to many potential users of these statistics. In Welsch and Peters (1978) and Belsley, Kuh, and Welsch (1980), we modified (18) somewhat and used

$$\frac{x_i^T b - x_i^T b(i)}{s(i)\sqrt{h_i}} = \left(\frac{h_i}{1-h_i}\right)^{\frac{1}{2}} \frac{e_i}{s(i)\sqrt{1-h_i}} \tag{23}$$

as our measure of influence. The numerator is the difference between $\hat{y}_i$ and the prediction, $x_i^T b(i)$. The denominator is just an estimate of the standard deviation of the fit, $\sigma\sqrt{h_i}$. Although (23) is easier to interpret, I feel that (18) is a better overall diagnostic tool because it gives more emphasis to the leverage, h_i. For additional discussion of this point see Pregibon (1979).

The quantity

$$e_i/s(i)\sqrt{1-h_i} \tag{24}$$

is just the scaled predicted residual (Hoaglin and Welsch, 1978)

$$\frac{e_i/(1-h_i)}{s(i)(1+x_i(X^T(i)X(i))^{-1}x_i^T)^{\frac{1}{2}}} \tag{25}$$

and is also the t-statistic for testing if a dummy variable with a single one in the ith position should be added to the model (X-matrix). I have called this the studentized residual while others call $e_i/s\sqrt{1 - h_i}$ by the same name. Since, under the Gaussian model,

$$E(e_i^2/s^2(1 - h_i)) = 1 \tag{26}$$

it has always seemed to me that this should be called the standardized residual.

The distinction between s and s(i) is important in other contexts as well. Cook (1977) chose to measure influence by

$$\frac{(b - b(i))^T X^T X(b - b(i))}{s^2 p}$$

$$= \frac{1}{p} \left(\frac{h_i}{1 - h_i}\right) \frac{e_i^2}{s^2(1 - h_i)} . \tag{27}$$

This is similar to (23), but there are important differences. For example, when all of the observations but one lie on a line, (27) can give potentially confusing information since it may indicate that some observations on the line are more influential than the one observation not on the line. This is counterintuitive since deletion of this one observation leads to a perfect fit.

To see this let $x_1 = x_4 = 0$, $x_2 = \delta \neq 0$, $x_3 = -\delta$; $y_1 = 0$, $y_2 = \delta$, $y_3 = -\delta$, $y_4 = t \neq 0$ and fit a simple linear regression model with intercept. If we let $I_i(s(i))$ and $I_i^2(s)$ denote (23) and (27) respectively, then $I_4^2(s) < I_i^2(s)$, $i = 2, 3$ while $|I_4(s(i))| > |I_i(s(i))| i = 1,2,3$. The idea behind this example is based on geometric arguments contained in Dempster and Green (1981). Atkinson (1981) also discusses (23) and (27) in a somewhat different context.

To complete our discussion of approximate influence functions, it is profitable to consider one other approach. Attach a weight w_i to the ith observation only and let $b(w_i)$ be the corresponding weighted least-squares estimate. Then from Belsley, Kuh and Welsch (1980) we have

$$\frac{\partial b(w_i)}{\partial w_i} = (X^TX)^{-1}x_ie_i/(1 - (1 - w_i)h_i)^2.$$

When $w_i = 0$ this is essentially (14). It is easily seen from (20) that $b - b(i)$ lies between this derivative evaluated at $w_i = 0$ and at $w_i = 1$.

IV. CUT-OFFS FOR DIAGNOSTICS

Once a measure of influence has been defined, the need arises for a cut-off to determine a highly influential observation. There are at least four different approaches; balance, exploration, statistical analogies, and formal tests.

One of the reasons for measuring influence is to see if there is severe imbalance in the influence of the individual observations. Ideally, the influence of each observation would be about the same with some allowance for stochastic variation. From (11) we have that

$$\underset{x,y}{E}\ (\Omega(x,y,F,T)^T V^{-1}\Omega(x,y,F,T)) = p \tag{28}$$

or

$$\frac{1}{n-1}\sum_{k\neq i}\Omega(x_k,y_k,F_{n-1},T)^T[V(i)]^{-1}\Omega(x_k,y_k,F_{n-1},T) \cong p$$

and we can use this as a rough guide for our finite sample approximations to (10) based on the square of (18). Thus we would consider there to be unbalanced influence when

$$\left(\frac{h_i}{1 - h_i}\right)\frac{e_i^2}{s^2(i)(1 - h_i)^2} \geq ap/(n-1) \tag{29}$$

for some constant a > 1. A reasonable value for a can be found by considering the case p = 1 where the X-matrix is just a column of ones. The left-hand side of (29) reduces to

$$\frac{e_i^2}{s^2(i)(1 - \frac{1}{n})^2 (n - 1)} .$$

Now $e_i^2/s^2(i)(1 - \frac{1}{n})$ has an $F_{1,n-2}$ distribution under the Gaussian model. When n > 15 the one percent point is less than 9 and therefore a $\cong$ 9 is a fairly conservative choice. The same cut-offs would be approximately applicable to (23). These cut-offs are closely related to the theory of bounded-influence regression. Further details are contained in Krasker and Welsch (1979) and Krasker, Kuh, and Welsch (1981).

The exploratory approach considers an influence measure as a batch of n numbers and used the techniques of exploratory data analysis including stem-and-leaf plots, box plots, and transformations to symmetry (logs often help here), in order to identify unusual observations. The features most noticed are gaps between groups of observations with influence of approximately the same magnitude. As a screening process to nominate observations for further consideration, the exploratory approach is hard to beat. Its real advantage is that it forces one to think and avoids the automatic use of cut-offs.

The fact that

$$\left(\frac{h_i}{1 - h_i}\right)^{\frac{1}{2}} \frac{e_i}{s(i)\sqrt{(1 - h_i)}} = \frac{x_i^T b - x_i^T b(i)}{s(i)\sqrt{h_i}} \tag{30}$$

has been cited as one reason for adopting (23) instead of (18). It also allows us to argue that, say, a change of one standard error between $x_i^T b$ and $x_i^T b(i)$ is worthy of note. We are <u>not</u> comparing $x_i^T b - x_i^T b(i)$ to its own standard error; but instead to the standard error of the fit. When a value of one or two is considered large, we are by analogy comparing the ratio to a t-statistic, even though it does not have a t-distribution.

Cook (1977) has used a similar idea, noting that (27) can be compared to the formula for finding a joint confidence interval for β and hence to the F-statistic. This does not mean that (27) has an F-distribution, just that a statistical analogy is being used. Weisberg (1980) states that a useful cut-off is 1 which translates to

$$\left(\frac{h_i}{1 - h_i}\right) \frac{e_i^2}{s^2(1 - h_i)} > p \tag{31}$$

and seems very large in view of (29) and the discussion which follows it.

V. INFLUENTIAL SUBSETS

Up until now we have concentrated on measuring the influence of adding one more observation. What happens if we add more than one observation? The formal theory of influence functions (Huber, 1981) implies that if G is a distribution function not too far from F in some metric, then

$$T(G) - T(F) \cong \int \Omega(x,y,F,T)\,dG(x,y). \tag{32}$$

For example,

$$G(x,y) = (1 - \varepsilon)F(x,y) + \frac{\varepsilon}{2}\,\delta_{x_1,y_1} + \frac{\varepsilon}{2}\,\delta_{x_2,y_2}. \tag{33}$$

Diagnostic work requires finite sample analogs to (32). A natural way to start is to consider adding d observations (denoted by D) to the remaining n-d observations. Then for least-squares we can evaluate (5) with F replaced by F_{n-d} to obtain

$$\Omega(x_i,y_i,F_{n-d},b(D)) = (n - d)(X^T(D)X(D))^{-1}x_i(y_i - x_i^T b(D)) \tag{34}$$

where $i \varepsilon D$, X(D) consists of the (n-d) remaining rows of X and b(D) is the least-squares estimate obtained without using the observations whose indices are in D. The variance approximation (17) becomes

$$V(D) = (n - d)s^2(D)(X^T(D)X(D))^{-1} \tag{35}$$

and using (10) we have

$$[(n - d)x_i^T(X^T(D)X(D))^{-1}x_i(y_i - x_i^T b(D))^2/s^2(D)]^{\frac{1}{2}} \tag{36}$$

as a (scaled) measure of influence for observations with $i \varepsilon D$.

If our goal is to identify influential observations and not necessarily influential subsets, then we need to find those observations with $i \varepsilon D$ where (36) exceeds some appropriate cut-off. Of course, doing this for all possible subsets D is expensive and some procedures for reducing the computing burden are contained in Belsley, Kuh and Welsch (1980).

Alternatively, asymptotic influence functions can be used to identify influential observations by finding bounded-influence regression estimates and a weight for each observation. Low weights indicate influential observations. This effective low-cost alternative to the all possible subsets approach is discussed in Krasker and Welsch (1979).

We can convert (36) into a way to identify influential subsets by declaring that a subset is influential if, for at least one index $i \epsilon D$, (36) exceeds an appropriate cut-off. However, this does not provide a very useful overall measure of the influence of one subset compared to that of another subset. We would find such a measure by counting the number of influential points in D, adding up or taking the product of the influences, finding a mean or geometric mean influence for a subset, or using the largest influence of a point in D as the measure of influence for D. All of these have merit in certain situations.

Recent work on subset diagnostics in Belsley, Kuh, and Welsch (1980) and Cook and Weisberg (1980) has emphasized other approaches. These are based on (32) and measure the influence of a subset by adding up the individual influences. To use (32) correctly we must assign a weight say, $c_i \geq 0$, to each added observation (we used 1/2 in (33)) such that $\sum_{i \epsilon D} c_i = 1$. In what follows we will use $c_i = 1/d$. Unequal weights can be useful, but computational costs are very high.

All this suggests that we should measure the influence of a subset by computing the sum

$$I_D = \frac{1}{d} \sum_{i \varepsilon D} (n - d)[X^T(D)X(D)]^{-1} x_i(y_i - x_i^T b(D)) \tag{37}$$

and reduce this to a scalar by using (10) and (35) to obtain

$$Q_D^2 = I_D^T [X^T(D)X(D)] I_D / s^2(D)(n - d). \tag{38}$$

Let

$$U = X_D(X^T(D)X(D))^{-1} X_D^T$$

$$H_D = X_D(X^T X)^{-1} X_D^T$$

where a D subscript denotes the matrix with just the rows (or rows and columns in the case of H_D) indexed by D. Using results obtained by Bingham (1977) we have

$$Q_D^2 = e_D^T U(I + U)^2 e_D / s^2(D) d^2(n - d)$$

$$= e_D^T H_D (I - H_D)^{-3} e_D / s^2(D) d^2(n - d). \tag{39}$$

where e is the vector of least-squares residuals. Note that Q_D strongly resembles (18) and in fact equals (18) when the only index in D is i.

While (38) is an appealing overall measure of influence for a subset D, it has not, as yet, seen wide use. Cook and Weisberg (1980) suggest

$$(b - b(D))^T X^T X(b - b(D))/ps^2$$

$$= e_D^T U(I + U) e_D / ps^2$$

$$= e_D^T H_D (I - H_D)^{-2} e_D / ps^2 \tag{40}$$

as a direct generalization of (27). Welsch and Peters (1978) suggested appropriately scaled forms of

$$(b - b(D))^T X^T(D) X(D) (b - b(D))$$

$$= e_D^T U e_D$$

$$= e_D^T H_D (I - H_D)^{-1} e_D \tag{41}$$

which generalizes (22) and can be computed very efficiently using the methods described in Belsley, Kuh, and Welsch (1980).

The cost of many of these procedures depends, in part, on what we are interested in. If we want to be able to find influential subsets and rank order them by an influence measure such as (38), (40), or (41), then the cost will be high. However, we will be able to see if the groupings found have meaning. For example, they may be nearby census tracts or related by some characteristic not in the model. This approach has been emphasized in Belsley, Kuh, and Welsch (1980).

Cook and Weisberg (1980) favor giving up some of this information in return for computational speed. They argue that in many situations we are interested primarily in finding subsets of observations that contain no observations that have been declared individually influential. Hence we can exclude all subsets from computation that contain such an observation or observations.

In implementing this procedure, Cook and Weisberg suggest an algorithm that may overlook some of the subsets they are trying to find. For example, assume that when the subset size is one, observations a, b, c, and d are not declared influential. When the subset size is two assume (a,c), (a,d), and (c,d) are not influential but (a,b) is declared to be influential. Cook and Weisberg declare a _and_ b to be individually influential because they are part of an influential subset. This means that when the subset size is three, subset (a, c, d) will not be considered as one that contains no individually influential observations. However, this subset may well be influential and c or d would not be declared individually influential.

The problem is that without a way to measure individual observation influence in a subset such as (36), it is difficult to declare an individual point influential just because it is part of an influential subset. It is easy to modify the Cook and Weisberg procedure to overcome this. For example, with subsets of size three, only exclude subsets with both a _and_ b present since they form an influential subset. Sets with just a or b (such as (a, c, d)) would have their influence computed. In effect, we will be looking for influential subsets that contain no previously declared influential _subsets_. This method will, of course, increase computational costs. Until more comparisons can be made, we cannot be sure if the compromises suggested in the Cook and Weisberg algorithm are worth the loss of information.

Subset diagnostics is an area of active research and no one method will serve every need. Lower computing costs will allow the creation of large amounts of potentially useful information. Finding a way to plot or look at this information remains the greatest challenge.

VI. CUT-OFFS FOR SUBSETS

Determining cut-offs for influential subsets is a complex matter. When (36) or a similar measure is used for individual observations within a subset, the cut-offs discussed in part IV are reasonable since we are dealing with the influence of a single observation and not of the whole subset, D.

When examining the influence of a whole subset, some adjustment may be necessary for the size of the subset since, as more observations are set aside, we expect generally greater influence. When adding influences, such as in (37) it is natural to divide by the size of the subset, d, and use the criteria suggested in part IV for Q_D. Cook and Weisberg (1980) use a cut-off that is independent of d because of the analogy of their approach to the F-statistic for confidence regions.

There is no theoretically correct answer to the determination of cut-offs. However, the testing approach of Dempster and Green (1981) applied to subsets may provide useful clues. Simulated examples comparing the above ideas will be reported on at a later date.

VII. FINAL COMMENTS

We have introduced a large number of potential diagnostics in this paper. At present we favor (18) and (38) as good overall diagnostics because they reflect the influence of observations on the coefficients, shape of the covariance matrix, and scale. All three are important for inference. If a more specialized procedure is required, (23) and its generalization [(40) with ps^2 replaced by $d^2s^2(D)$] reflect changes in coefficients and scale, but not shape. Finally, the measures introduced by Cook, (27) and (40) reflect only changes in coefficients.

For single observations, computational costs are not significant and choices among diagnostics may be made on the basis of utility and experience. When subsets are involved, the best diagnostics may be too expensive to obtain and compromises are often needed. Many of these compromises will be problem specific (size of data set, types of subsets suspected, etc.) and we can make no general recommendations at this time.

ACKNOWLEDGMENTS

Helpful conversations with Stephen C. Peters, Alexander M. Samarov, and William S. Krasker and the typing of Joanne Sorrentino are gratefully acknowledged.

REFERENCES

Atkinson, A.C. (1981). "Two Graphical Displays for Outlying and Influential Observations in Regression," Biometrika 68, 13-20.

Belsley, D.A., Kuh, E. and Welsch, R.E. (1980). Regression Diagnostics: Identifying Influential Data and Sources of Collinearity. Wiley, New York.

Bingham, C. (1977). "Some Identities Useful in the Analysis of Residuals from Linear Regression," Technical Report 300, School of Statistics, University of Minnesota.

Cook, R.D. (1977). "Detection of Influential Observations in Linear Regression," Technometrics 19, 15-18.

Cook, R.D. and Weisberg, S. (1980). "Characterizations of an Empirical Influence Function for Detecting Influential Cases in Regression," Technometrics 22, 495-508.

Dempster, A.P. and Gasko-Green, M. (1981). "New Tools for Residual Analysis," Ann. of Statist. (in press).

Hampel, F.R. (1968). Contributions to the Theory of Robust Estimation. Ph.D. Thesis, Univ. of Calif., Berkeley.

Hampel, F.R. (1974). "The Influence Curve and its Role in Robust Estimation," J. Amer. Statisti. Assoc. 62, 1179-1186.

Hoaglin, D.C. and Welsch, R.E. (1978). "The Hat Matrix in Regression and ANOVA," Amer. Statistician 32, 17-22.

Huber, P.J. (1981), Robust Statistics. Wiley, New York.

Krasker, W.S., Kuh, E., and Welsch, R.E. (1981). "Estimation for Dirty Data and Flawed Models." Handbook of Econometrics, edited by Z. Griliches and M.D. Intrilligator. North-Holland (in press).

Krasker, W.S. and Welsch, R.E. (1979). "Efficient Bounded-Influence Regression Estimation Using Alternative Definitions of Sensitivity," Technical Report No. 3, M.I.T. Center for Computational Research in Economics and Management Science, Cambridge, Mass. (To appear J. Amer. Stat. Assoc.)

Pregibon, D. (1979). Data Analytic Methods for Generalized Linear Models, Ph.D. Thesis, Univ. of Toronto.

Pregibon, D. (1981). "Logistic Regression Diagnostics," Ann. of Statist. 9, 705-724.

Rao, C.R. (1973). Linear Statistical Inference and its Applications. Wiley, New York.

Tukey, J.W. (1970). Exploratory Data Analysis, Mimeograph Preliminary Edition.

Weisberg, S. (1980). Applied Linear Regression. Wiley, New York.

Welsch, R.E. and Peters, S.C. (1978). "Finding Influential Subsets of Data in Regression Models." Proceedings of the Eleventh Interface Symposium on Computer Science and Statistics, edited by A.R. Gallant and T.M. Gerig, Institute of Statistics, North Carolina State University Raleigh, N.C. 240-244.

THE USE AND INTERPRETATION OF ROBUST ANALYSIS OF VARIANCE

Joseph W. McKean

Department of Mathematics
Western Michigan University
Kalamazoo, Michigan

Ronald M. Schrader

Department of Mathematics and Statistics
The University of New Mexico
Albuquerque, New Mexico

I. INTRODUCTION

Classical analysis of variance (ANOVA) is one of the principal statistical research tools in many scientific disciplines, its position probably more assured now than over thirty years ago when Eisenhart (1947) made a similar observation. The primary reason for its popularity is that it provides a summary in convenient tabular form of complex patterns in data. This advantage accrues whether one regards the associated test statistics and p-values in an inferential or in a purely discriptive manner.

As a simple example, consider a repeated measures problem encountered by one of the authors. Sixty subjects were divided into three treatment groups of twenty each and the value of the dependent variable y was measured on each subject at ten different time intervals. Notable immediately in the ANOVA were large treatment $\times$ time and subject $\times$ time interaction

ISBN 0-12-438180-4

effects. Plots of y against time for each subject revealed that some subjects failed to follow the general pattern, but as a rule values of y decreased more rapidly for subjects in one treatment group than in the other two groups.

The analysis of variance served two important purposes in this problem. First, it directed the experimentor's attention to a distinct pattern which he needed to investigate further. This initial screening can be much more important in still more complicated designs. Second, it allowed a simple summary of the patterns in the data, a summary which had among other advantages that it was sufficiently compact to meet the space requirements of a professional journal. Sixty different scatter plots were required to yield as much information.

The need for a robust analysis of variance follows from the advantages of classical ANOVA. Many experimentors truly like the classical ANOVA and will rely heavily upon it if no comparable robust procedure is available. Since entries in a classical ANOVA table are functions of least squares estimates of regression coefficients, and since least squares estimates are by now notorious for their lack of robustness, analogous procedures based upon robust estimates are needed.

Such procedures are available. McKean and Hettmansperger (1976) developed one based upon the R-estimates of Jaeckel (1972). Schrader and Hettmansperger (1980) developed another based upon the M-estimates of Huber (1973). Both procedures are as general and widely applicable as classical ANOVA, and provide a complete and analogous compact summary of data. We show in Sections 2 and 3 that the robust methods have as rich and flexible a structure as classical ANOVA.

To see the similarity between a robust ANOVA and the classical procedure, as well as the advantage of the robust method, consider the following example from John (1978). Twenty-seven observations were collected in a 3^{4-1} factorial design, the factors designated as A, B, C, and D. Three degrees of freedom from the four factor interaction were used as defining contrasts $(I = A B^2 C^2 D^2)$. Parameters fit were the constant, the linear and quadratic contrasts for each effect, and the linear × linear interactions.

From a plot of least squares residuals and an outlier test it is evident that one value is an outlier; the experimentor was unable to explain the anomaly. John's analysis consists of replacing the corresponding observation with a missing value estimate. The analysis of variance shifts from linear A being the only apparent effect to several effects being in evidence, notably a $B \times C$ interaction term. Clearly presence of a single outlier seriously limits the ability of classical ANOVA to find any patterns. Parts of the ANOVA tables from the original and outlier adjusted least squares analyses appear in Table I.

The need for ANOVA is more evident in this example than in the last example. The data appear in a sparse manner in this fractional factorial design, hindering the use of many graphical devices. The large effects detected in the ANOVA suggest areas for closer examination.

One goal of robust analysis is to provide automatic detection and adjustment for outliers. Also in Table I is part of a robust ANOVA based upon Huber's M-estimates, as discussed in Schrader and Hettmansperger (1980). This analysis is explained

TABLE I. ANOVA for Example by John (1978)

Source of Variation	Least Squares F (Original Data)	Least Squares F (Outlier Adjusted)	Robust F_M (Original Data)
Linear A	5.23[a]	13.80[b]	9.75[b]
Quadratic A	0.79	9.82[b]	4.71[a]
Linear B	0.25	5.18[a]	1.52
Quadratic B	0.42	3.63	2.14
Linear C	0.97	2.56	1.69
Quadratic C	0.14	5.27[a]	1.57
Linear D	0.19	0.54	0.01
Quadratic D	0.11	0.08	0.00
Linear A × Linear B	0.06	0.17	0.00
Linear A × Linear C	0.32	0.86	1.07
Linear B × Linear C	2.68	7.09[a]	6.50[a]

[a] Significant at 5 percent level

[b] Significant at 1 percent level

fully in Sections 2 and 3, but for now it sufficies to know that the entries F_M are compared to usual F crictical values.

While not in perfect agreement with John's analysis, the M-ANOVA has detected many of the same patterns discovered by John. In Section 4 we present Monte Carlo evidence of the validity of the robust ANOVA in this problem.

It might be noted that while median polishing (Tukey, 1977) does not solve precisely the same problem, it leads to conclusions substantially in agreement with the robust ANOVA. It is necessary to confound other effects with effects due to D to reach such a conclusion, however, due to the sparsity of data.

We turn now to a more mathematical description of robust ANOVA. This is done to demonstrate that the rich structure of the classical procedure is present as well in our proposed robust procedure. Use and interpretation of the robust procedure is greatly enhanced by the presence of so much structure.

II. GEOMETRY OF ROBUST ANOVA

Assume we have formulated the problem of interest in terms of a linear model. Let Y denote the $n \times 1$ vector of observations, the response variable, and consider the model

$$Y = X\beta + e \tag{2.1}$$

where X is an $n \times p$ matrix, β is a $p \times 1$ vector of unknown parameters, and e is an $n \times 1$ vector of random errors. The columns of the design matrix X may be composed of dummy variables or regression variables (covariates) or both. The vector β is a set of natural parameters for the problem. Let r be the rank of X. In general X is of non-full column rank, that is, $r < p$. Let Ω be the r-dimensional subspace of R^n spanned by the columns of X. Then except for random error Y would be in Ω.

Since prediction of Y and the distance between Y and subspaces of R^n are fundamental to estimation and testing, they will be discussed first. After first reviewing least squares, we will present two methods of robust prediction, R- and M-. These ideas generalize the R- (Jaeckel (1972)) and M- (Huber (1973)) estimation of regression coefficients.

A. Prediction and Minimum Distances

Based on the model (2.1) we want to determine the "best" prediction of Y from the subspace Ω. Best depends on the criteria being used least squares, R-, or M-estimation.

The best prediction of Y based on least squares is the projection, $\hat{Y}_L$, of Y onto Ω; that is $\hat{Y}_L$ satisfies

$$||Y-\hat{Y}_L||_L^2 = \min ||Y-\eta||_L^2, \quad \eta \in \Omega$$

with the squared norm defined by

$$||u||_L^2 = \Sigma u_i^{\,2}, \quad u \in R^n. \tag{2.2}$$

For convenience denote the squared Euclidean distance between Y and Ω by $D_L(\Omega) = ||Y-\hat{Y}_L||_L^2$. This is the usual sum of squared errors, SSE, for least squares. We will refer to it as the least squares minimum dispersion for the model (2.1).

R-best predictions are obtained by considering a norm other than the Euclidean. Let $0 \leq a(1) \leq a(2) \leq \ldots \leq a(n)$ be a sequence of scores on the first n-positive integers. Define the function

$$||u||_R = \Sigma a(R|u_i|)\,|u_i|, \quad u \in R^n \tag{2.3}$$

where $R|u_i|$ is the rank of $|u_i|$ among $|u_1|, \ldots, |u_n|$. These scores are often called signed rank scores. The L_1-norm can be obtained by taking the scores $a(i) \equiv 1$. Another score function that is often used is the Wilcoxon where $a(i) = i$. McKean and Schrader (1980) show that $\|\cdot\|_R$ is a norm on R^n; hence there exists a vector $\hat{Y}_R \in \Omega$ which satisfies

$$\|Y-\hat{Y}_R\|_R = \min \|Y-\eta\|_R, \quad \eta \in \Omega.$$

Call $\hat{Y}_R$ an R-best prediction of Y and denote the R-minimum distance between Y and Ω by $D_R(\Omega) = \|Y-\hat{Y}_R\|_R$. We will refer to this as the R-minimum dispersion for model (2.1). Note that since $\|\cdot\|_R$ is a norm no initial estimate of scale is needed to obtain an R-best prediction.

M-best predictions of Y are obtained by considering the function

$$\nu(u,v) = \Sigma\rho((u_i-v_i)/\hat{\sigma}), \quad u, v \in R^n \tag{2.4}$$

where $\hat{\sigma}$ is an initial estimate of scale and ρ is a function on R which satisfies

(i) $\rho(x) \geq 0$, $\rho(x)$ is non-decreasing for $x \geq 0$

(ii) $\rho(0) = 0$

(iii) $\rho(-x) = \rho(x)$

(iv) $\psi(x) = \rho'(x)$ exists and is continuous at all but a finite number of x.

Two examples of such ρ-functions are:

$$\rho_H(x) = \begin{cases} \frac{1}{2}x^2 & \text{if } |x| \leq k \\ k|x| - k^2/2 & \text{if } |x| > k \end{cases}; \quad \rho_A(x) = \begin{cases} c^2[1-\cos(x/c)] & \text{if } |x| \leq \pi c \\ 2c^2 & \text{if } |x| > \pi c \end{cases}$$

The first is given by Huber (1973) and the second by Andrews (1974). Let $\hat{Y}_M$ be a vector in Ω which satisfies

$$\nu(Y,\hat{Y}_M) = \min \nu(Y,\eta), \quad \eta \in \Omega.$$

Such a vector does exist and in general a function g can be defined so that $g(\nu(u,v))$ defines a metric on R^n. Let $D_M(\Omega) = \nu(Y,\hat{Y}_M)$ be the M-minimum dispersion for the model (2.1).

B. Computation of Robust Predictions

All of the best predictions, and, hence minimum distances defined above are computationally feasible. For least squares, there are several ways of obtaining projections. One of the most numerically stable methods is to obtain a QR-decomposition of X. This yields orthonormal bases for both Ω and $\Omega^{\perp}$ as a product of r-Householder transformations. A set of efficient algorithms for QR-decompositions and a discussion of their uses can be found in LINPACK (Dongarra, et. al., 1979).

Once a QR-decomposition of X has been obtained, M-best predictions can be found by using Huber's (1977) proposal 2. This algorithm operates iteratively on residuals, and as shown in the Appendix, for each iteration the increment to the residuals is simply a projection on Ω which can be obtained efficiently by using the QR-decomposition. R-best predictions are obtained similarly by employing a QR-decomposition of X and the Gauss-Newton algorithm discussed in McKean and Hettmansperger (1978).

C. Estimation

The vector β contains the parameters of interest in an experiment; hence, given a best prediction $\hat{Y} \in \Omega$ of Y by one of the above methods we would like to solve the system of equations

$$X\hat{\beta} = \hat{Y} \tag{2.5}$$

for $\hat{\beta}$. In general X is not of full column rank so there is an infinite number of solutions. Usually we are not interested in estimating β itself but certain linear functions of β. Those that can be uniquely determined we term estimable:

Definition. For $\lambda \in R^p$, the function $\lambda'\beta$ is said to be an estimable function of β, if for any $\eta \in \Omega$ the value of $\lambda'\beta$ is the same for all solutions β of the system $X\beta = \eta$. In particular if $\lambda'\beta$ is estimable then the value of $\lambda'\hat{\beta}$ is unique for all $\hat{\beta}$ satisfying (2.5).

This definition of estimability is similar to Rao's (1973, p. 223) first definition. Note that it avoids assuming that expectations exist, an assumption which is undesirable from a robust point of view. As in the classical theory an equivalent formulation is more convenient:

Theorem. $\lambda'\beta$ is estimable if and only if λ' lies in the row space of X.

The proof can be found in McKean and Schrader (1980).

Suppose we want to estimate $H\beta$ where H is a $q \times p$ matrix whose row space is contained in the row space of X. Computationally the least squares estimates follow easily from the QR-decomposition of X, see Golub and Styan (1973, p. 270). Robust estimates are obtained similarly by replacing the projection of Y with a robust prediction.

D. Tests of General Linear Hypotheses

Hypotheses of interest about β can usually be expressed in the form

$$H_O : H\beta = 0 \quad \text{versus} \quad H_A : H\beta \neq 0 \tag{2.6}$$

where H is a $q \times p$ matrix such that $H\beta$ is a set of linearly independent estimable functions.

As in Scheffé (1959, p. 33) the problem can be cast in a geometric framework. Under the full model (2.1), the column space of the design matrix is Ω with dimension r. Under H_O the range of the design matrix is restricted to a subspace

ω of Ω of dimension $r-q$. This is called the reduced model. Let $D_L(\omega)$, $D_R(\omega)$, $D_M(\omega)$ denote the distances as defined in Section 2.1 between the vector Y and the subspace ω.

The test procedure for all the methods discussed above is a comparison of the distances between the observations Y and each of the subspaces Ω and ω. This is illustrated in the Figure where distances and predictions could be any of the methods. Natural test statistics for H_0 are the difference in distances $D(\omega) - D(\Omega)$, large values of which would indicate H_A.

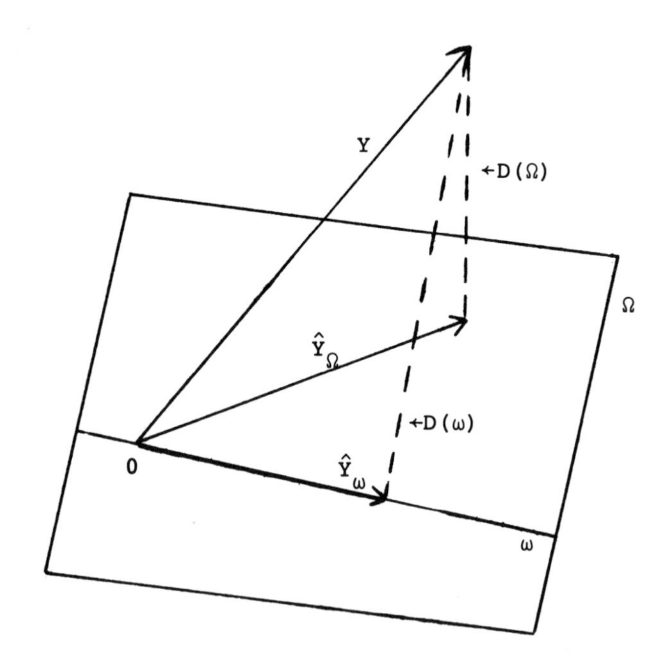

FIGURE. **Predictions and minimum distances of Y from the from the subspaces Ω and ω.**

The difference in distances is the reduction in minimum dispersion as we pass from the reduced to the full model. It corresponds to the classical reduction in sums of squares due to H_o. Suitable standardizing constants for the robust reductions in dispersion are discussed in the next section. The resulting tests can then be summarized in a robust analysis of variance table quite similar to the classical table, see Schrader and McKean (1977) for discussion and examples.

Computationally these tests require best predictions of Y from ω. As noted in Appendix B a basis for ω can be obtained easily by using a QR-decomposition of the matrix H'.

III. ASYMPTOTIC THEORY

A. Standardization of Tests of Section 2.

In order to use the natural test statistics, the difference in distances, suitable standardizing constants are required. We will ultimately appeal to asymptotic theory which we motivate by likelihood ratio tests.

Consider the hypothetical case where the density of the errors, $f(x)$, is proportional to $\exp\{-\rho^*(x/\sigma)\}$ (for non-decreasing ψ-functions this is plausible). Next suppose we choose the associated correct criterion, $\nu^*(u,v) = \Sigma\rho^*((u_i-v_i)/\sigma)$, to obtain our best predictors. Let $D_*(\Omega)$ denote the minimum distance. Then the likelihood ratio test statistic for the hypotheses (2.6) is

$$\Lambda = \exp\{-D_*(\omega)\}/\exp\{-D_*(\Omega)\}.$$

Under these assumptions $-2\log\Lambda = 2(D_*(\omega)-D_*(\Omega))$ converges in distribution to a central χ^2-random variable with q degrees of freedom; see Rao (1973, p. 419). The distribution is exact if the errors are normal and the criterion is least squares.

Generally, of course, the density of the errors is unknown. Assume we choose to use M-best predictions determined with the function ρ as given in (2.4). Define the parameter τ by

$$\tau = E[\psi^2(e/\sigma)]/E[\psi'(e/\sigma)].$$

Then under H_o and suitable regularity conditions Schrader and Hettmansperger (1980) show that $2\tau^{-1}[D_M(\omega)-D_M(\Omega)]$ converges in distribution to a $\chi^2(q)$ random variable. This suggests as a robust F-test of H_o the statistic

$$F_M = (D_M(\omega)-D_M(\Omega))/q\hat{\lambda}_M , \tag{3.1}$$

where $\lambda_M = \tau/2$ is estimated with full model residuals r_i by

$$\hat{\lambda}_M = \frac{1}{2}(N-r)^{-1}\Sigma\psi^2(r_i/\hat{\sigma})/\{N^{-1}\Sigma\psi'(r_i/\hat{\sigma})\};$$

see Schrader and Hettmansperger (1980).

Next assume we use R-best predictions, (2.3), with scores $a(i)$ generated by $a(i) = \phi(i/(n+1))$ where ϕ is a non-negative, non-decreasing function defined on the interval $(0,1)$. For example $\phi(u) \equiv 1$ generates the L_1-norm and $\phi(u) = u$ generates the Wilcoxon scores. Let $A^2 = \int\phi^2(u)du$ and $\phi_F(u) = -f'(F^{-1}(u))/f(F^{-1}(u))$ where f is the density of the errors. Define the paramters

$$\gamma = \int_o^1\phi(u)\phi_F((u+1)/2)\,du \quad \text{and} \quad \lambda_R = A^2/2\gamma.$$

Then following the development of McKean and Hettmansperger (1976) and Kraft and vanEeden (1972), under H_o the random variable $\lambda_R^{-1}[D_R(\omega)-D_R(\Omega)]$ has an asymptotic $\chi^2(q)$ distribution. This suggests as a test statistic

$$F_R = [D_R(\omega)-D_R(\Omega)]/(q\hat{\lambda}_R) \tag{3.2}$$

where estimation of λ_R is discussed in McKean and Hettmansperger (1978).

Behavior of the robust statistics F_R and F_M under alternatives is discussed in McKean and Hettmansperger (1976) and Schrader and Hettmansperger (1980). Under H_A both tests are consistent. Under a sequence of contiguous alternatives both $q \cdot F_R$ and $q \cdot F_M$ have non-central $\chi^2(q)$ distributions. The ratio of the resulting non-centrality parameters to that of least squares yields the usual asymptotic relative efficiences associated with parameter estimates.

B. Alternative Analyses

Asymptotic distribution theory for the robust estimates, Huber (1973) and Jaeckel (1972), suggest another test statistic for the hypotheses (2.6). Under suitable regularity conditions $H\hat{\beta}$ has an asymptotic $N(H\beta, \kappa^2 H(X'X)^- H')$ distribution where

$$\kappa^2 = \begin{cases} \sigma^2 E[\psi^2(e/\sigma)]/E[\psi'(e/\sigma)]^2 & \text{for M-estimates} \\ A^2/\gamma^2 & \text{for R-estimates} \end{cases}$$

and $(X'X)^-$ is a generalized inverse of $X'X$. Thus the asymptotic theory suggests test statistics based on a quadratic form in $H\hat{\beta}$:

$$F_Q = (H\hat{\beta})'[H(X'X)^- H']^{-1}(H\hat{\beta})/q\hat{\kappa}^2, \tag{3.3}$$

where $H\hat{\beta}$ is a robust estimate for the full model (2.1) and $\hat{\kappa}^2$ is the associated estimate of κ^2. The matrix inversion needed to compute F_Q can be performed efficiently using QR-decompositions; see Golub and Styan (1973). The analysis using the test statistic F_Q is equivalent to the analysis using pseudoobservations proposed by Bickel (1976, p. 167).

Preliminary Monte Carlo study, Schrader and Hettmansperger (1980), indicates that the test statistic F_Q, with the usual F-critical values, tends to become liberal with heavy tailed errors. Perhaps further work will uncover better standardizing constants and degrees of freedom corrections.

IV. Monte Carlo Study of Robust ANOVA

In order to verify that the robust ANOVA reported in Table I is valid, one thousand repetitions of the same 3^{4-1} design reported in Table I were independently simulated with $N(0,1)$ (standard normal) and CN2549 (N(0,1) contaminated with probability .25 by N(0,49)) errors. The Huber proposal 2 scale estimate (Huber, 1977) was employed, as well as the Huber function

$$\rho\,(x) = \begin{cases} \frac{1}{2}\,x^2 & |x| \leq 1 \\ |x| - \frac{1}{2} & |x| > 1 \end{cases} .$$

Note that the "bend" for Huber's function was chosen to be 1.0. It is the experience of the authors that with Huber proposal 2 scale estimation it is necessary to let the function $\rho(x)$ bend somewhere near 1.0 in order to gain much power with heavy tailed errors.

The simulation results appear in Table II. This design is rather removed from those handled by asymptotic theory, having twenty-seven observations and twelve parameters. In this context it is rather encouraging that significance levels of F_M adhere so closely to nominal levels.

TABLE II. Emperical α-Levels (Percents) from 1000-Simulation of John's Model

Hyp	Nominal α	Normal Errors: Least Squares F	Normal Errors: Robust F_M	CN2549 Errors: Least Squares F	CN2549 Errors: Robust F_M
	01	011	023	005	006
Linear A	05	060	074	042	061
	10	110	121	099	120
	01	008	012	005	004
Quadratic A	05	046	062	040	042
	10	105	118	095	098
	01	006	014	006	010
Linear B	05	045	054	056	060
	10	093	110	104	121
	01	012	018	003	005
Quadratic B	05	054	063	036	042
	10	109	127	098	093
	01	003	007	010	015
Linear C	05	040	045	040	051
	10	084	094	087	106
	01	007	010	001	003
Quadratic C	05	047	055	036	033
	10	112	118	093	105
	01	008	008	007	008
Linear D	05	042	055	048	057
	10	096	109	109	104
	01	009	015	003	003
Quadratic D	05	048	056	031	041
	10	102	121	081	099
	01	010	016	013	011
Lin A × Lin B	05	051	066	051	054
	10	093	109	102	125
	01	006	009	012	016
Lin A × Lin C	05	044	058	046	055
	10	089	114	095	105
	01	007	009	010	013
Lin B × Lin C	05	050	060	050	062
	10	090	107	092	124

V. APPENDIX

A. Algorithm for M-best Predictions

The following algorithm is a non-full rank version of Huber's (1977, p. 39) H-algorithm. It employs a QR-decomposition of the design matrix. Consider the general linear model (2.1) where X is a $n \times p$ matrix of rank r. Let Ω = range X. Write the QR-decomposition, Golub-Styan (1973, p. 267), of X as

$$X = Q \begin{bmatrix} R & S \\ 0 & 0 \end{bmatrix} \Pi'$$

where Q is an o.n. basis for R^n, R is a $r \times r$ upper triangular matrix of rank r, and Π is a $p \times p$ permutation matrix. Further, $Q = [Q_1 \vdots Q_2]$ where Q_1 is an $n \times r$ matrix whose columns form an orthonormal basis for Ω. Note that the projection onto Ω is then $Q_1 Q_1'$. Projections and transformations of the form $Q_1'Z$ can be efficiently obtained by using algorithms found in LINPACK, Dongara et. al. (1979). As in Huber let $\chi(x) = x\psi(x) - \rho(x)$ and $a = (n-r)E[\chi(U)]$, where U has a standard normal distribution.

Let $r_i^{(0)}$, $i = 1,\ldots,n$, be initial residuals and $\hat{\sigma}^{(0)}$ be an initial estimate of scale. Following Huber the algorithm is:

1. Put $m = 0$.
2. Form the new value for σ
$$\left(\hat{\sigma}^{(m+1)}\right)^2 = \frac{1}{a} \Sigma\chi\left(r_i^{(m)}/\hat{\sigma}^{(m)}\right)\left(\hat{\sigma}^{(m)}\right)^2.$$
3. Form the vector Z where
$$z_i = \hat{\sigma}^{(m+1)}\psi\left(r_i^{(m)}/\hat{\sigma}^{(m+1)}\right).$$
4. Let $\hat{\tau} = Q_1'Z$.

5. Form new residuals

$$r^{(m+1)} = r^{(m)} - qQ_1Q_1'Z$$

where q is a relaxation factor $(0 < q < 2)$.

6. Stop iterating if

$$|q\hat{\tau}_i| < \varepsilon\, \hat{\sigma}^{(m+1)} \quad i = 1,\ldots,r$$

and $|\hat{\sigma}^{(m+1)} - \hat{\sigma}^{(m)}| < \varepsilon\, \hat{\sigma}^{(m+1)}$.

Otherwise put $m = m + 1$ and return to step 2.

The output is a set of M-residuals. The algorithm converges for non-decreasing ψ-functions, see Huber (1977, p. 40). For redescending ψ-functions we suggest using initial residuals from one of Huber's ρ-functions.

B. Reduced Model Design Matrix

Consider the hypothesis (2.6). Note $H\beta = 0$ if and only if $\beta \in \ker H = (\text{range } H')^{\perp}$. Form a QR-decomposition of the matrix H',

$$H' = Q_H \begin{bmatrix} R_H \\ 0 \end{bmatrix}.$$

Then $Q_H = [Q_{H1} \vdots Q_{H2}]$ where Q_{H2} is an orthonormal basis for $\ker H$. Hence the reduced model design matrix is $U = XQ_{H2}$.

REFERENCES

Andrews, D.F. (1974). A Robust method for multiple linear regression. *Technometrics 16*, 523-31.

Bickel, P.J. (1976). Another look at robustness: A review of reviews and some new developments (reply to discussant). *Scand. J. Statist. 3*, 167.

Dongarra, J.J., Bunch, J.R., Moler, C.B., and Stewart, G.W. (1979). "LINPACK Users' Guide." Philadelphia: Society for Industrial and Applied Mathematics.

Eisenhart, C. (1947). The assumptions underlying the analysis of variance. *Biometrics 3*, 1-21.

Golub, G.H. and Styan, G.P. (1973). Numerical computations for univariate linear models. *J. Statist. Comput. Simul. 2*, 253-274.

Huber, P.J. (1973). Robust regression: Asymptotics, conjectures, and Monte Carlo. *Ann. Statist. 1*, 799-821.

Huber, P.J. (1977). "Robust Statistical Procedures". Philadelphia: Society for Industrial and Applied Mathematics.
Jaeckel, L.A. (1972). Estimating regression coefficients by minimizing the dispersion of residuals. *Ann. Math. Statist.* *43*, 1449-1458.
John, J.A. (1978). Outliers in factorial experiments. *Appl. Statist.* *27*, 111-119.
Kraft, C.H. and vanEeden, C. (1972). Linearized rank estimates for the general linear hypothesis. *Ann. Math. Statist.* *43*, 42-57.
McKean, J.W. and Hettmansperger, T.P. (1976). Tests of hypotheses based on ranks in the general linear model. *Comm. in Statist. A5*, 693-709.
McKean, J.W. and Hettmansperger, T.P. (1978). A robust analysis of the general linear model based on one step R-estimates. *Biometrika* *65*, 571-579.
McKean, J.W. and Schrader, R.M. (1980). The geometry of robust procedures in linear models. *J.R. Statist. Soc.B.*, (To appear).
Rao, C.R. (1973). "Linear Statistical Inference and its Applications", 2nd ed. New York: John Wiley.
Scheffé, H. (1959). "The Analysis of Variance". New York: John Wiley.
Schrader, R.M. and Hettmansperger, T.P. (1980). Robust analysis based upon a likelihood criterion. *Biometrika* *67*, 93-101.
Schrader, R.M. and McKean, J.W. (1977). Robust analysis of variance. *Comm. Statist. A.* *6*, 879-894.
Tukey, J.W. (1977). "Exploratory Data Analysis". Reading, Mass.: Addison-Wesley.

Research of both authors partially supported by U.S. Army grant DAAG 29-79-C-002.

AUTHOR INDEX

SUBJECT INDEX

E

S

W

X

Y